Cooperative Extension

MW01715575

FARM RESCUE

Responding to Incidents and Emergencies in Agricultural Settings

Revised by

William E. Field
Professor
Department of Agricultural and Biological Engineering
Purdue University

Authors of the 1986 edition of *Farm Accident Rescue* (on which this revision is based)

L. Dale Baker • Product Safety Engineer
J. I. Case Corporation • Hinsdale, Illinois

William E. Field • Professor • Department of Agricultural and Biological Engineering • Purdue University

Rollin Schnieder • Department of Agricultural Engineering • University of Nebraska

Clair W. Young • Professor Emeritus
The Ohio State University

Dennis J. Murphy • Professor and Extension Safety Specialist • Department of Agricultural and Biological Engineering • The Pennsylvania State University

Natural Resource, Agriculture, and Engineering Service (NRAES)
Cooperative Extension
PO Box 4557
Ithaca, New York 14852-4557

NRAES–10
Revised August 1999

Revised April 1986 • Revised April 1982 • First Edition 1980

ISBN 0-935817-32-8

Library of Congress Cataloging-in-Publication Data

Farm rescue : responding to incidents and emergencies in agricultural settings. -- Rev. / William E. Field.
p. cm. -- (NRAES ; 10)
Revision of the 1986 ed. of Farm accident rescue.
Includes bibliographical references.
ISBN 0-935817-32-8 (pbk.)
1. Farmers--Diseases--Handbooks, manuals, etc. 2. Agriculture--Accidents--Handbooks, manuals, etc. 3. Medical emergencies--Handbooks, manuals, etc. I. Field, William E., 1949– . II. Farm accident rescue. III. Series: NRAES (Series) ; 10.
RC965.A5F37 1999
616.02'5'08863--dc21 98-53184
CIP

Natural Resource, Agriculture, and Engineering Service (NRAES)
Cooperative Extension
PO Box 4557
Ithaca, New York 14852-4557
Phone: (607) 255-7654
Fax: (607) 254-8770
E-mail: NRAES@CORNELL.EDU
Web site: WWW.NRAES.ORG

Contents

Figures

Chapter 1: Overview of Farm Rescue

Chapter 2: Agricultural Equipment Injuries and Rescue

Chapter 3: Farm Structure Emergencies

Chapter 4: Farm Chemical Exposures, Fires, and Spills

Chapter 5: Farm Animal Incidents

Appendix A: Rescue Tool Inventory Guidelines

Appendix B: Agricultural Operator Control Symbols

Tables

Chapter 1

Overview of Farm Rescue

Purpose of This Handbook

Responding to farm-related injuries, entrapments, fires, and chemical incidents can be both demanding and hazardous for rural emergency response personnel, including firefighters, emergency medical personnel, rescue squads, and law enforcement officers. Many of these individuals respond as part of volunteer units that may not be as well equipped or as highly trained as their urban counterparts. As a result, each year there are tragic cases of emergency rescue personnel being killed or seriously injured while using inappropriate procedures in farm-related rescue attempts.

The purpose of this handbook is to help familiarize emergency medical and rescue personnel with basic principles and procedures for responding to agricultural emergencies. The most common types of incidents are described, and potential hazards to victim and responder are clearly identified.

This handbook does not discuss medical procedures that are covered by other areas of rescue training. Nor do the authors suggest that the material addresses every situation that might be encountered in a farm-related incident. Rather, this book is designed to supplement additional training and experience in order to enable rescuers to approach a farm rescue with more confidence and a greater awareness of the risks involved.

Overview

Rural emergency personnel should be prepared for a variety of agricultural emergencies, such as tractor overturns, machinery entanglements, grain bin entrapments, exposure to toxic gases, injuries involving agricultural chemicals or pesticides, and incidents concerning farm animals (figure 1.1). The following are characteristics generally associated with farm-related injuries and emergencies:

- unusual structures
- confined spaces
- high-rise structures
- heavy machinery
- livestock
- chemicals
- toxic environments
- remote locations
- exposure to the elements

Responding to farm equipment- and facility-related injuries is, in some obvious ways, similar to responding to

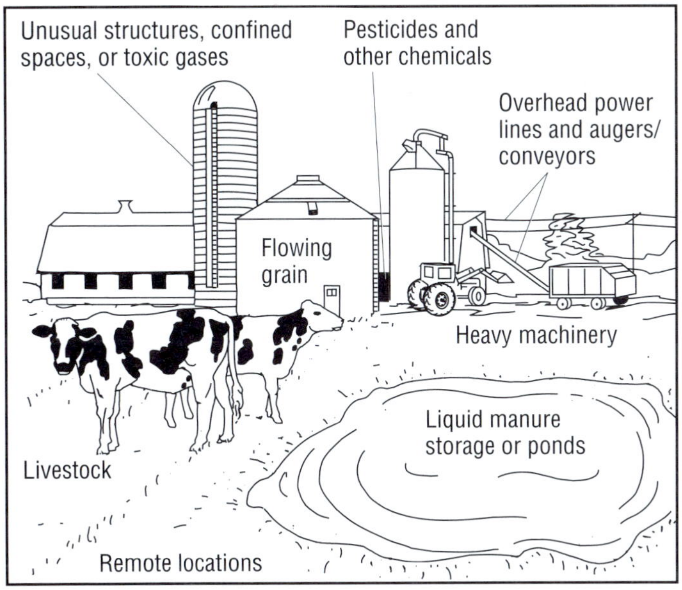

Figure 1.1
Potential farm hazards

motor vehicle crashes. Other factors, however—such as poor accessibility to the scene, the ruggedness of agricultural equipment, and the occasional need for expertise on disassembly or extrication procedures—make responding to farm-related incidents different enough to require specialized training and preparation (figure 1.2). In addition, some older or lighter-weight rescue equipment, such as hydraulic rescue tools and metal-cutting saws, which are effective in freeing victims trapped in automobile crashes, may not work on agricultural machinery. Generally, on-farm rescue procedures will likely require more planning time, and disassembly of machinery will probably take much longer than would a similar operation after a motor vehicle crash.

Obtaining additional help and equipment will also take longer when greater distances are involved. Farm injuries may occur in far-from-ideal locations—in the center of a muddy field, for example. The surroundings may preclude the use of vehicle-mounted lights, air systems, or electrically powered rescue tools. In these cases, it is advisable to begin arranging alternative sources of power and light—such as tractors from neighboring farms—even while rescuers are on their way to the scene. If it is determined that the victim will need to be transported to a hospital, an air ambulance helicopter should be called in early in the rescue, if possible.

One of the most significant issues faced by responding emergency personnel is the likelihood that they will be unfamiliar with agricultural settings, equipment, facilities, and processes. Few people have a farm background anymore. This unfamiliarity has lead to mistakes and additional injuries during rescue efforts on farms. Gaining a better understanding of what to expect when responding to a farm emergency is important, not only for the welfare of the patient but also for the safety of those providing assistance.

The following chapters cover the most common farm emergencies in which rescue, extrication, or other emergency response may be needed. However, numerous variables make it impossible to cover every emergency situation. The following incidents are discussed in detail.

Agricultural equipment injuries and emergencies (chapter 2)

- tractor overturns
- power takeoff driveline (PTO) entanglements
- hydraulically operated equipment failures
- mixer wagon and self-unloading wagon entanglements
- auger and elevator entanglements
- electrocutions
- injuries and emergencies involving harvesting equipment—balers, combines, and corn pickers

Incidents involving farm structures (chapter 3)

- flowing-grain entrapments
- exposures to silo gas, injuries involving silo unloaders, and silo and hay mow fires
- emergencies involving manure storage sites and facilities

Farm chemical exposures, fires, and spills (chapter 4)

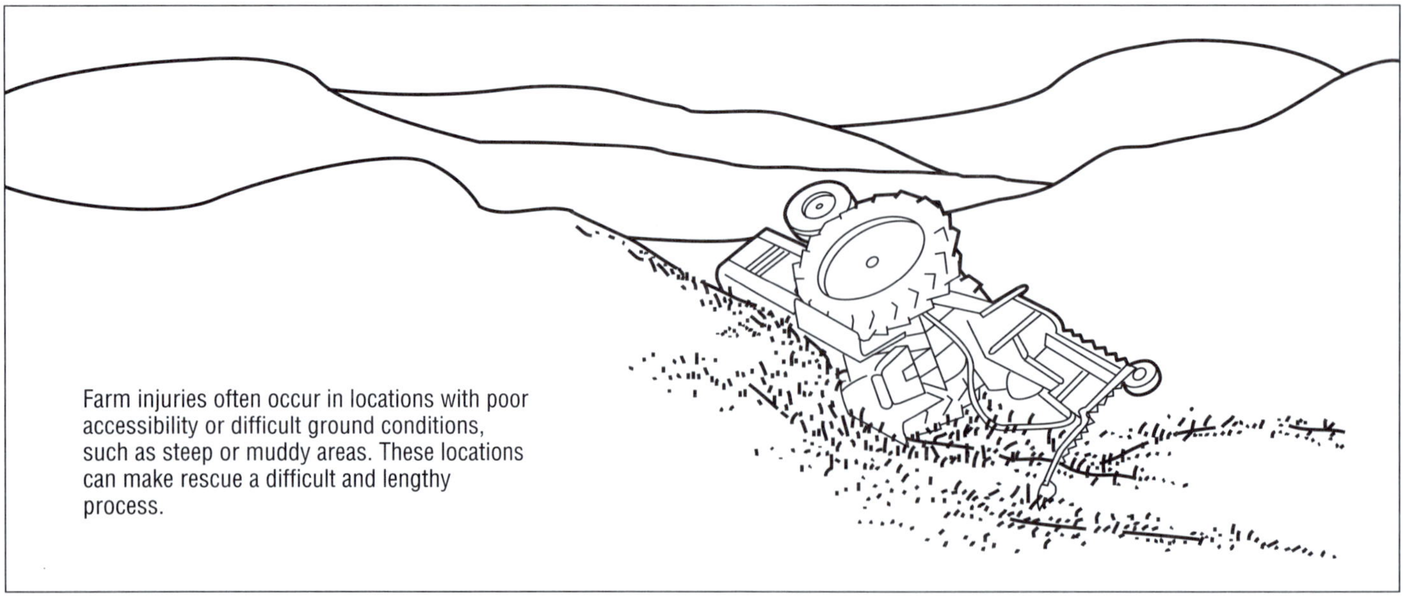

Figure 1.2
Farm-related injuries can occur in difficult locations.

- anhydrous ammonia exposures
- agricultural chemical fires
- pesticide exposures and spills

Farm animal incidents (chapter 5)

Discussions in this book include true accounts of farm incidents that reflect the rescue procedures covered. These real-life scenarios vividly portray a wide range of farm-related injuries and emergencies, while summing up the information and the coordination strategies that can lay the foundation for an effective rescue. They also illustrate the difference between the ideal and the actual, showing how real-life rescues may not conform to the conditions presented in this book or in other rural rescue publications.

Organization of the Rescue Team

Especially during the first few minutes after the emergency unit arrives at the scene, good leadership is essential to putting into practice a safe and successful rescue.

The good leader
- analyzes the incident and plans the rescue before beginning
- assigns such tasks as establishing the command post and safety area
- decides on a course of action
- coordinates the rescue
- knows the limitations of the rescue unit and its equipment
- does not hesitate to call for additional help.

Most rural rescue units train more than one leader, or "incident commander," since the same individual cannot be available for every emergency. This practice is in accordance with the Incident Command System, a method of structuring the management of the incident scene (figure 1.3). Every incident requires an incident commander to oversee the entire operation without participating personally in the extrication. The safety officer is the individual responsible for the overall safety of the incident scene. In order to ensure a safe scene, the safety officer needs to have enough authority to stop any potentially harmful action he or she sees. The rescue sector commander oversees the rescue component of the scene—the actions taken toward getting the victim out. The emergency medical services (EMS) sector commander monitors the patient's condition and oversees medical attention. The rescue sector and EMS sector commanders need to work together to continuously assess the situation and communicate with one another. This structure allows for the most efficient and consistent response and victim care.

General Rescue Procedures

Three Steps to Stabilization for Safety

Every incident is unique; but the primary concern in any rescue is the safety of the victim, bystander, and rescuer. Before beginning patient care, the rescue unit should make the scene as safe as possible for everybody. This is done by following the "three fundamental *S*s," or steps to stabilization for safety (figure 1.4):

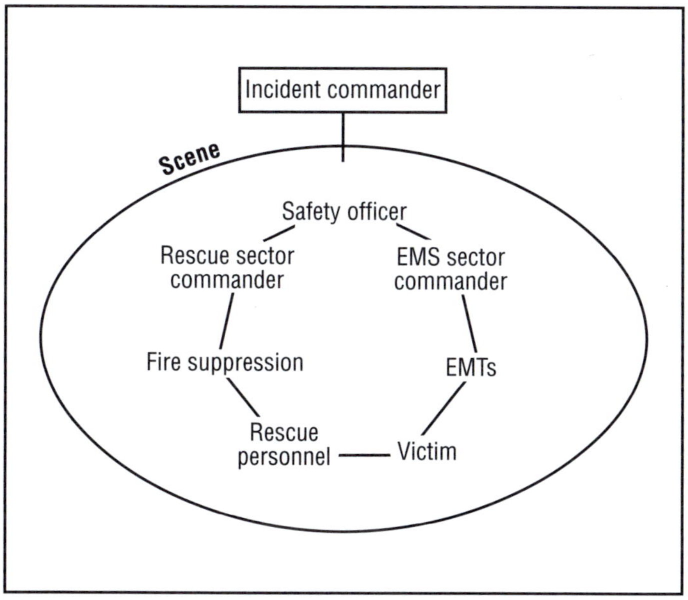

Figure 1.3
The Incident Command System

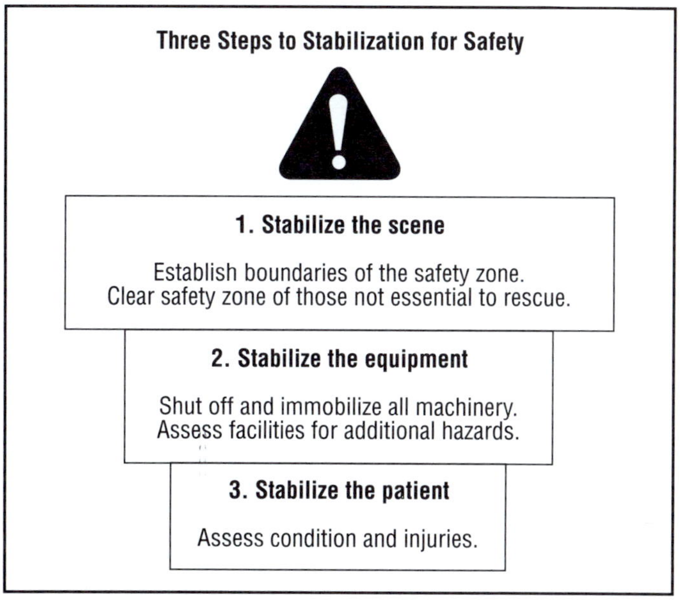

Figure 1.4
Three steps to stabilization for safety

1. Stabilize the scene (assess hazards; establish boundaries of the safety zone; and clear zone of anyone not essential to the rescue).

2. Stabilize the equipment (shut off and immobilize all machinery; assess facilities and structures for additional hazards).

3. Stabilize the patient.

1. Stabilize the Scene

The first step of a safe rescue is to establish a safe work area, or safety zone. This area gives rescuers freedom to work; however, it may contain potential hazards such as fire, toxic or flammable gases, downed electrical wires, unstable equipment, or structures that may collapse. Therefore, the safety zone should have no more rescuers than necessary and absolutely no bystanders. Trained rescue crew members and law enforcement personnel must identify and maintain this area. If gasoline, LP gas, anhydrous ammonia, or toxic chemicals are involved, the zone will need to be enlarged, and evacuations may be needed. One person should be assigned to give statements to the press, so as to prevent conflicting reports. Another member of the rescue team should be assigned to communicate with the patient's family members, coworkers, or neighbors. Even though they want to be helpful, concerned bystanders at the scene may interfere with the rescue.

2. Stabilize the Equipment/Structures

Many types of equipment, structures, and facilities have the potential to cause injury, even when they are not operating. Heavy machine components may suddenly drop or shift if levers are inadvertently moved; augers and grain dryers may start unexpectedly due to automatic timers and preprogrammed operating procedures; confined spaces, such as silos or manure pits, can collect high levels of toxic gases or be deficient in oxygen. Although time is critical, rushing into a scene to carry out a rescue can lead to further injury for the victim or to injury or even death for rescuers. Before the rescue is conducted, equipment must be secured, electrical systems shut down and secured, and confined spaces tested for toxic or hazardous atmospheres. To ensure a safe rescue, assessing hazards and stabilizing equipment and facilities are absolute musts.

3. Stabilize the Patient

Once the scene has been secured and equipment and facility hazards stabilized, attention can be given to the patient. Respiration, signs of bleeding or shock, and evidence of other types of injuries such as spinal cord injury or loss of a body part (see "Preserving Amputated Tissue," the next section) should be assessed. One rescuer must remain with the patient at all times to provide psychological support. If the patient's family members cannot be of immediate help, they should be kept out of the safe work area. In all farm rescues, medical-control communication should be established.

If the patient has lost an appendage or a large amount of skin or other tissue, try to locate the tissue for possible reattachment. Follow the steps summarized below to properly preserve the tissue (figure 1.5).

Preserving Amputated Tissue

1. Wrap the appendage or tissue in a dry, sterile dressing or towel.

2. Place wrapped appendage or tissue in a clean plastic bag and seal it. Label it with the victim's name, the date, and the time.

3. Place package in a cooler with ice or cold packs for transport to the hospital. Do not allow the tissue to freeze, however.

Contacting Help

Extrication from Agricultural Equipment and Structures

When a victim is entangled in agricultural machinery, both medical and mechanical expertise are necessary in order to extricate him or her. Maintaining good communications with the hospital and other rescue resources is extremely

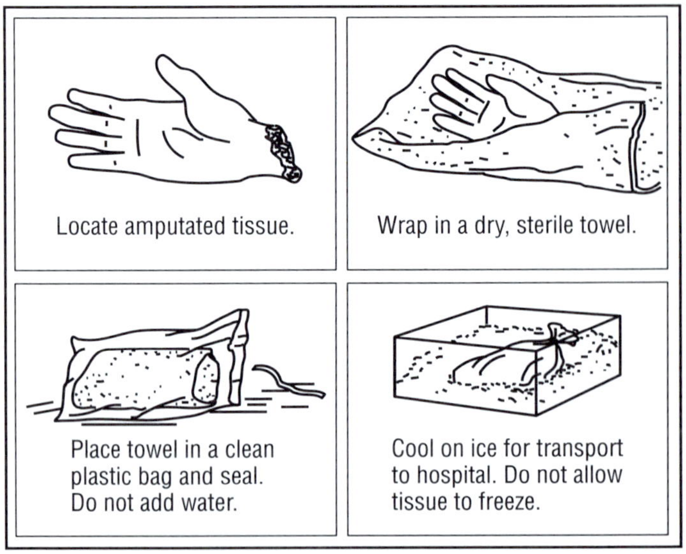

Figure 1.5
Caring for amputated tissue

important. For example, if it appears that an extrication will take considerable time, call for additional medical assistance from a resource such as a paramedic unit. In some cases, as when an emergency amputation is needed, it may be necessary to call a physician to the scene.

When travel distances are great, emergency personnel or those who can provide mechanical assistance need to be on the way as soon as possible if they are to arrive in time to help the victim. Call in technical assistance immediately if a complex piece of machinery needs to be disassembled, if the victim is in a confined space, or if a high-angle rescue is needed.

When communicating with a central dispatcher, provide the following information to pass on to other responding units:

- actual location of site (may not be the location from which the call for assistance was made)
- field conditions at the site (muddy, steep, snow-covered, etc.)
- type of equipment or structure involved
- victim's location at site
- victim's condition and nature of injuries

One step in preparing for a farm-related emergency is to fill in the telephone directory list of emergency resources that is provided on page 77 of this book. You may wish to post a copy of the completed list with your dispatcher and also keep a copy in each unit.

Do not overlook two local sources of assistance: neighboring farmers (who may be familiar with the equipment or facilities involved) and machinery dealers. If needed, they can help disassemble the machine. For the safety of the victim and the rescue team, however, the rescue unit should clearly be in charge of the procedure.

Keep accessible a current list of local farm machinery dealers (who are now scarce in many parts of the country). The best source of information on how to extricate the victim may be a dealer or service manager at a farm implement store that sells the same brand of machinery as that involved in an entanglement. If you cannot reach that dealership right away, other implement dealers often know the design of their competitors' machinery and usually can recommend extrication procedures.

Pesticide and Chemical Exposures, Fires, and Spills

For assistance in any pesticide or other chemical emergency, the most appropriate contacts will depend on community resources and state or local laws. Many counties now have hazardous materials (HAZMAT) response teams, which are prepared to safely handle chemical spills and fires. Frequently your local central dispatcher has ready access to relevant information about such questions as the toxicity of the chemical involved and whether or not a given chemical is flammable.

Consult the U.S. Department of Transportation's *North American Emergency Response Guidebook.* This publication contains information for first response to hazardous material incidents. (See suggested readings section, page 76, for more information on this book.)

You can also call CHEMTREC (1-800-424-9300), a nationwide service supported by chemical manufacturers to provide advice and assistance for handling such emergencies. CHEMTREC maintains a twenty-four-hour telephone hotline that is much like a referral service. Before you call, identify and make a note of the type of chemical involved or suspected, its brand name, and the amount involved.

Many state environmental protection/management agencies and agriculture departments can also respond to emergencies. In some cases, depending on the amount and type of chemical involved, reporting spills is required by law.

For medical assistance in treating poisoned patients, call the nearest poison control center.

Rescue Situations

The farm emergencies described in the following chapters are the most common that require extrication or other forms of rescue. Because of the many variables that can be involved, however, predicting the details of every emergency that could arise is impossible.

Disclaimer

To simplify information and illustrations, trade names have been used in this publication. No endorsement of named products is intended, nor is criticism implied of similar products that are not mentioned. NRAES makes no warranties, express or implied, concerning the accuracy, application, or use of the information mentioned or described in this publication. Further, nothing in this publication is intended as medical or legal advice. Competent professionals should be consulted on medical or legal issues, if any, arising from the use of the information contained in this publication.

Chapter 2

Agricultural Equipment Injuries and Rescue

Tractor Overturns

Tractor overturns contribute to more farm-related fatalities than any other source of farm injury. Even though rollover protective structures (ROPS) on newer tractors reduce the potential for injury during a tractor rollover, many older models lack these structures. To compound the hazard, these older tractors, which account for most fatal tractor injuries, are often used by less experienced operators, such as young people and part-time farmers. In many incidents, a farm implement was attached or being towed. This additional equipment can increase the severity of the injury and make the rescue more difficult.

Nearly 85 percent of tractor overturns are to the side (figure 2.1). Side overturns are caused primarily by hitting a raised object; dropping into a depression; operating with a raised front-end loader, which reduces stability; turning at too high a speed; or operating the tractor on a steep incline (figure 2.2).

In side overturns, the victim may sometimes be thrown clear of the tractor; in other cases, the victim may even have the opportunity to jump free. Severe injuries are possible even in these cases, however, due to the height and force of impact involved.

Rear overturns, in which the tractor revolves backward around the rear axle, occur less frequently but are more deadly than side overturns. The victim in a rear upset involving a tractor without an ROPS usually has little chance of escape and is often crushed between the seat and the ground. Rear overturns often occur as a result of improper hitching or traveling up a steep hill (figure 2.3).

Injuries

Injuries from tractor overturns vary considerably. Victims usually suffer crushing injuries to the head, chest, or pelvic areas. Torso injuries range from broken ribs to a crushed chest, with the possibility of a collapsed lung. Broken legs and arms and severe lacerations are also common. The extreme weight of the tractor resting on top of the victim

Figure 2.1
This tractor rolled down a hillside and trapped the operator underneath.

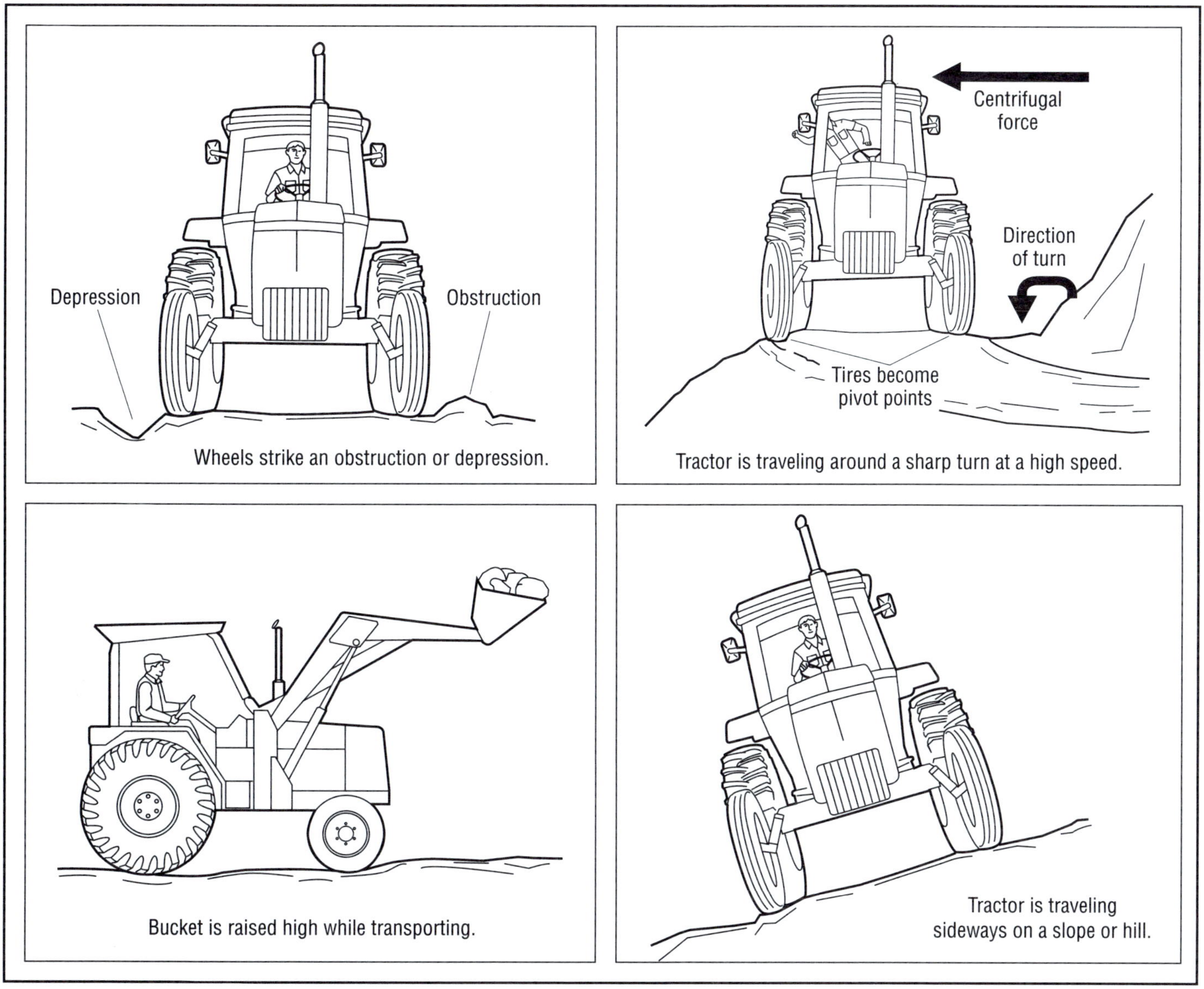

Figure 2.2
Reasons for side overturns

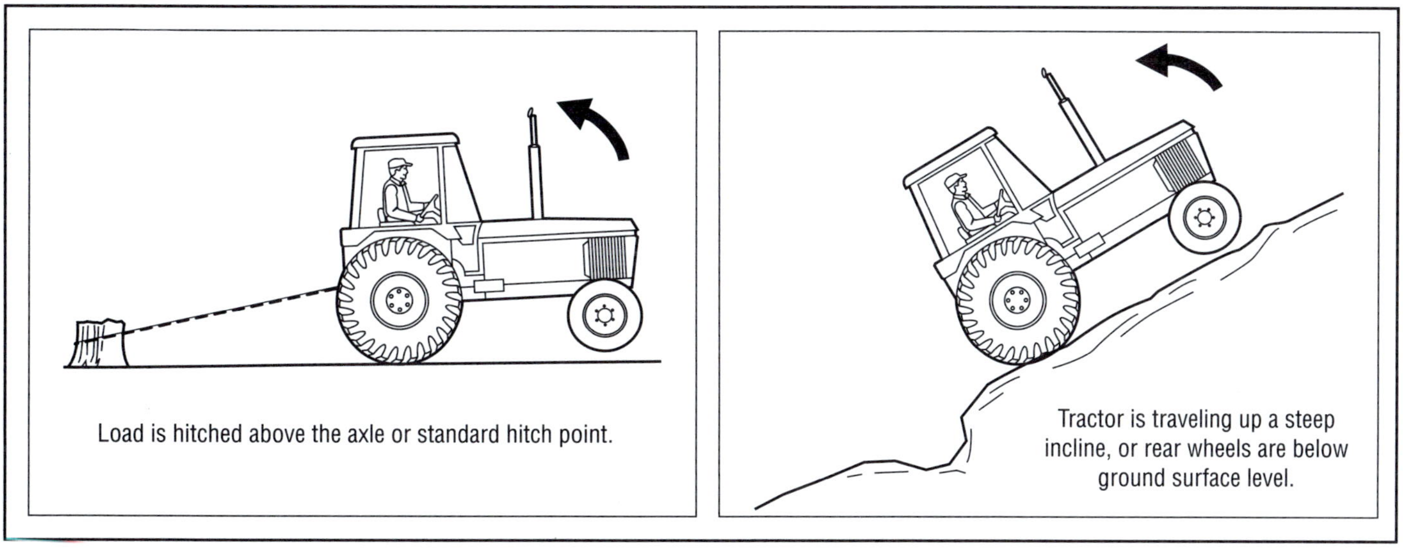

Figure 2.3
Reasons for rear overturns

and the extended time it may take to remove the tractor contribute to the high incidence of fatalities in these cases.

As in any crushing injury, the victim should not be moved more than is absolutely necessary until adequate support and medical personnel are present (figure 2.4). Since you may not know the full extent of the injuries, do not move the patient unless immediate danger exists. Movement may aggravate internal injuries and cause extensive internal bleeding or irreparable damage to the spinal cord. Before implementing the rescue, wait until the site, the tractor, and the patient have been stabilized and an adequate response team is in place. Following appropriate handling procedures for the patient is very important. In some cases, you should transport the patient in the position found, especially in cases of rear overturns, where the knees are sometimes crushed up against the chest. Permanent spinal cord injuries can be caused by improper handling after the initial injury.

Due to the severe bleeding often caused by blunt trauma associated with tractor overturn, the patient may experience hypovolemic shock. Indications of bleeding both external and internal should be addressed immediately and the patient monitored continuously for signs and symptoms of hypovolemic shock. Advanced life support (ALS) must be started at the scene and fluid replacement started in some situations. Other injuries associated with tractor upsets include burns from spilled engine coolant, hot transmission oil, hydraulic fluid, or battery acid. Flush acid burns immediately with large amounts of water. Remove contaminated clothing, if possible. **Caution:** If you suspect acid burns, give special attention to the victim's eyes. Flush with running water continuously for at least ten minutes.

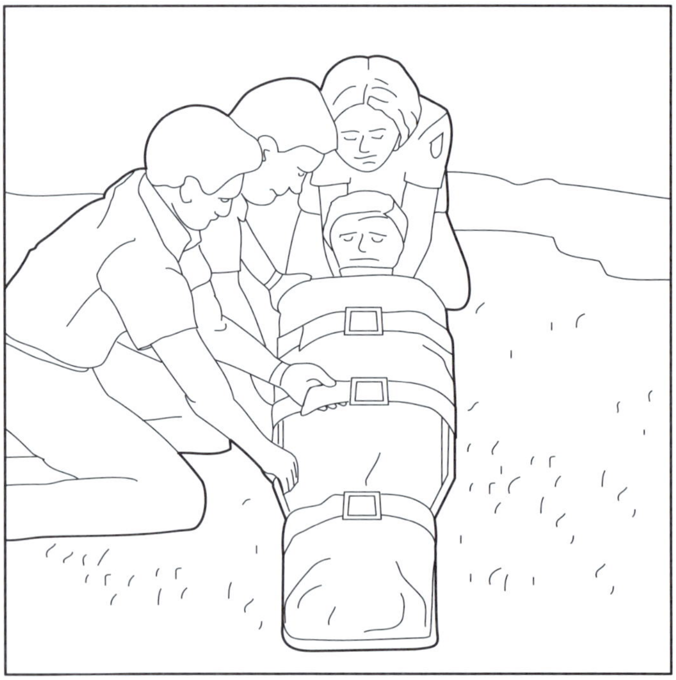

Figure 2.4
Properly securing and transporting a patient can prevent additional injuries, especially to the spinal cord.

Sometimes tractor tires are filled with a water and calcium chloride solution to increase the overall tractor weight. If a tire has ruptured, the victim might have been splashed with this mixture or another antifreeze solution. Such chemicals are strong irritants to the eyes, mucous membranes, and open wounds. Flush the contaminated area with water as soon as possible.

Rescue Procedures — Tractor Overturns

Essential Safety Precautions

1. Since many farm tractor overturns result in fatalities, entrapments, or serious injuries, it is important that appropriate emergency personnel respond to the scene. Responders to any farm tractor incident should include fire, rescue, and emergency medical services.

   Caution: Be alert for fire, and keep a charged 1½-inch or larger fire hose and both an ABC-type chemical fire extinguisher and a CO_2 fire extinguisher immediately at hand throughout the rescue.

2. Spilled gasoline, diesel fuel, or hydraulic oil may be a fire threat in any tractor overturn. Do not permit smoking around the site; and keep back curious onlookers, who might carelessly strike a match. Consider alternative methods to free a victim before using any cutting equipment that may generate a spark.

Shutting Off the Engine

Shut off the engine, even if it does not appear to be running. Rotation of the rear wheels or inadvertent contact with the starter solenoid could start the engine. A live ignition can also cause electrical arcing, which could ignite flammable liquids.

Gasoline-Powered Tractor Engines

If the engine is gasoline-powered (it would have spark plugs and a distributor), find the ignition switch on the control panel near the steering wheel and turn it off. Remove the key. Another approach is to disconnect the high-voltage wire running between the coil and distributor; this action will also shut off the engine. If gasoline is leaking, clear flammable vapors from the distributor area with a short discharge from a CO_2 extinguisher. Disconnect the battery to neutralize the electrical system, which, if damaged, could cause arcing.

Diesel Tractor Engines

If you are unfamiliar with tractors and engines, shutting off a diesel engine can be difficult. As with a gasoline-powered tractor, first try to locate a shutoff key on the control panel (figure 2.5). Turning the key to the "off" position will turn off the electrical system and **may** shut off an engine that is equipped with a solenoid-activated injector pump. Take out the key so that it cannot be unintentionally turned to the "on" position during rescue.

On some tractors, however, you may need to push or pull a fuel shutoff knob (usually red) into the "off" position and hold it in that position until the engine stops (see figure 2.6a). If the key or shutoff knob cannot be located or does not stop the engine when activated, locate the throttle lever, move it towards the idle position, and hold it in place (figures 2.6b, 2.6c, and 2.7). In some cases, this lever is also used to shut off the engine. Often, the throttle lever is directly above or beside the steering wheel. If none of the above procedures are successful, you can usually stop a diesel engine by manually operating the fuel shutoff lever

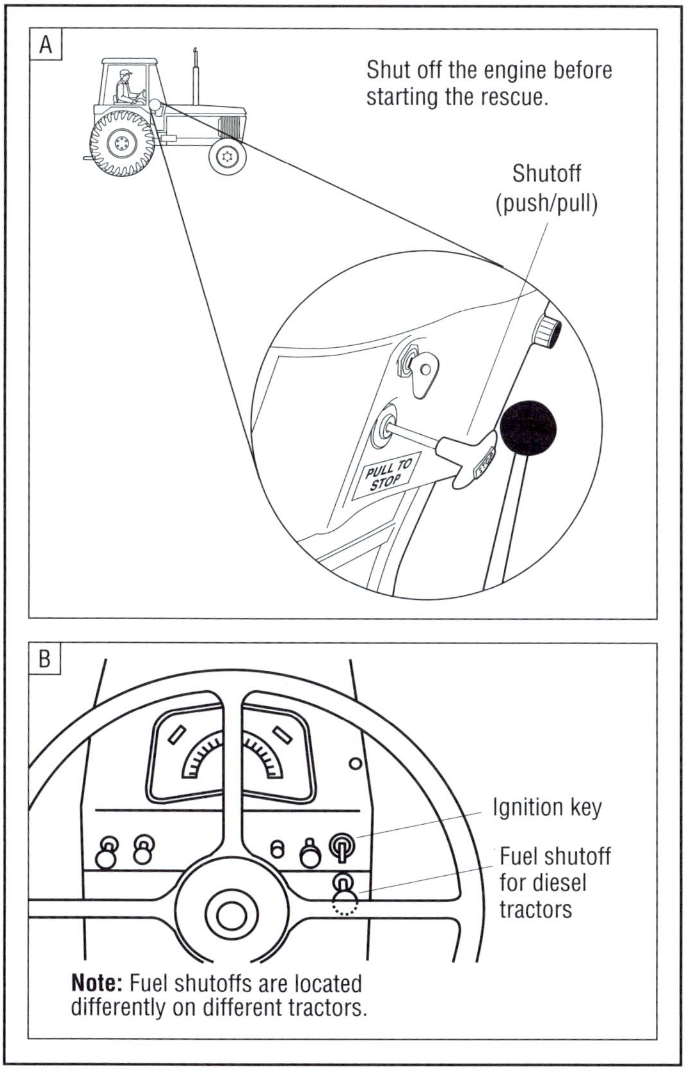

Figure 2.5
Typical shutoff locations for diesel tractors

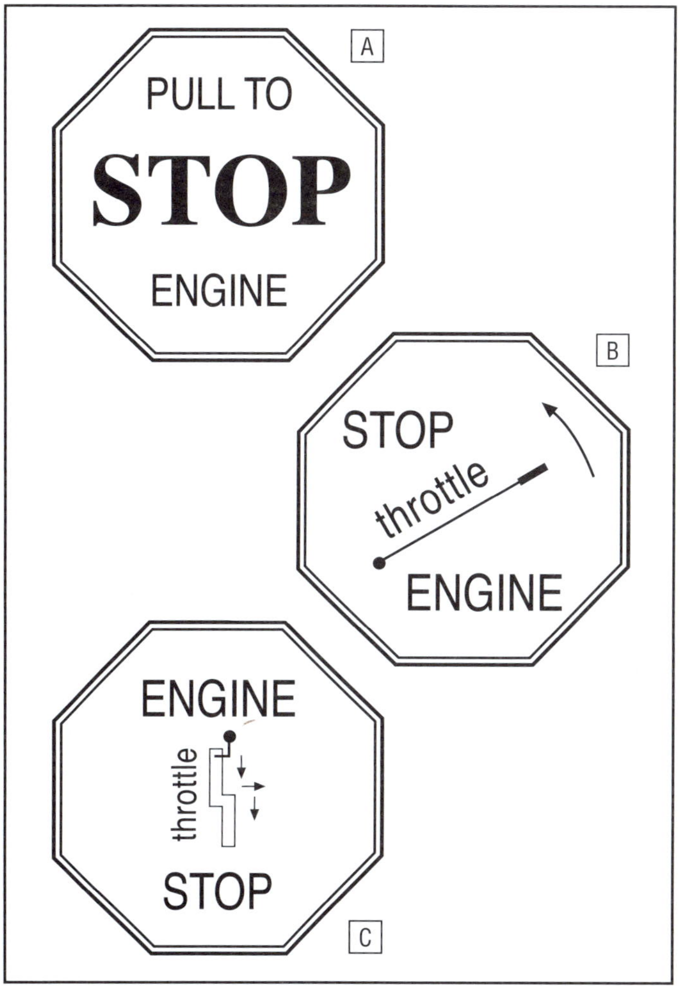

Figure 2.6
Engine stop decals. (a) "Pull to stop" engine decal. (b) Stop engine decal for clock-like throttle. (c) Engine stop decal for slot-like throttle.

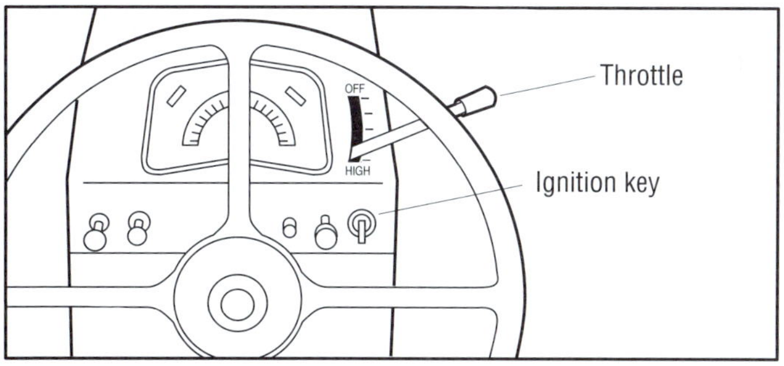

Figure 2.7
Location of the throttle

on the fuel injection pump (figure 2.8).

It may also be possible to shut off the engine by closing the fuel tank shutoff (figure 2.9). However, with this approach, the engine will run for several minutes while it uses the fuel left in the lines.

Last Resort for Shutting off a Diesel Engine

Shutting off the fuel supply as described above is the surest way of stopping a diesel engine. As a last resort when the fuel supply cannot be shut off, it may be possible to stop a diesel engine by discharging a CO_2 fire extinguisher (figure 2.10) or stuffing a rag or blanket into the air intake or air cleaner of the engine. A dry chemical extinguisher, however, may not plug the air cleaner sufficiently to stop the engine. In the case of air filters with dust-unloading valves, enough air may still reach the engine to keep it running.

Removing a Victim from under an Overturned Tractor

Digging under the Victim

The method used to remove the victim from under an overturned tractor will depend not only on the parts of the victim that are pinned but also on soil conditions and tractor size. If the ground is soft, it may be possible to dig him or her out. Digging under the victim can reduce the amount that the tractor must be lifted. Always stabilize the tractor so that digging will not cause it to tip or settle and further injure the victim or endanger rescuers. Stabilizing the tractor can be done with cribbing (figure 2.11), cabling, or a combination of both.

Cribbing

Once you begin removing the weight of the tractor from the victim, you must complete the rescue without setting the weight back down. Cribbing—adding support under the lifted machinery—is essential, whether you dig under

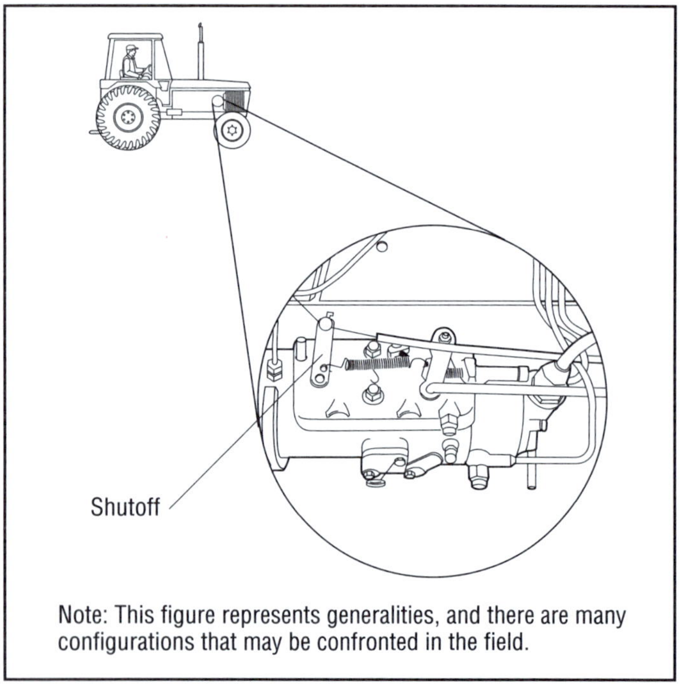

Figure 2.8
Diesel fuel injection pump shutoff

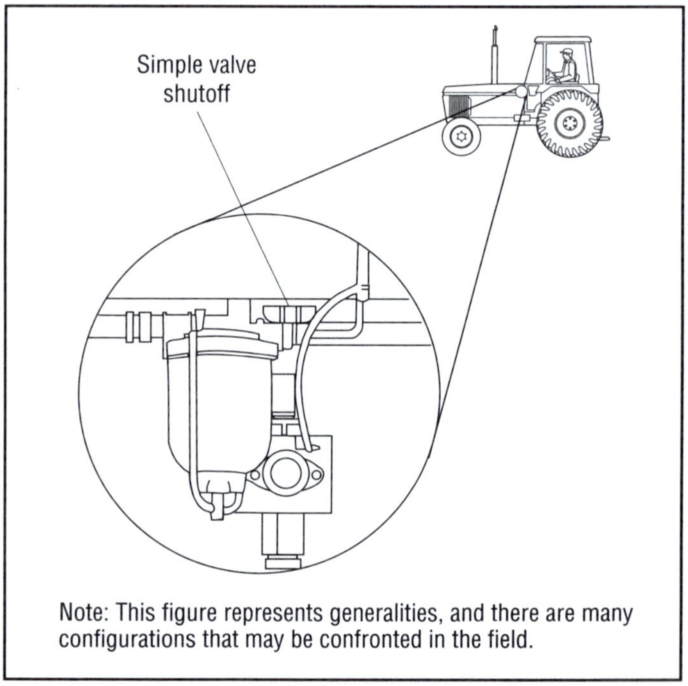

Figure 2.9
Fuel tank shutoff

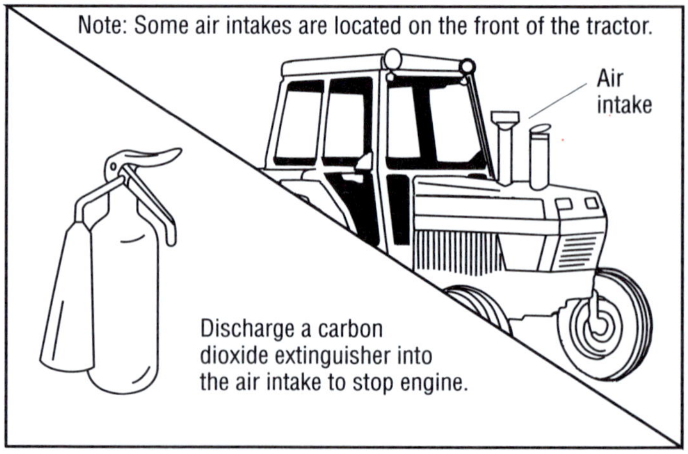

Figure 2.10
Stopping an engine by using a CO_2 extinguisher

a tractor or lift the tractor to free a victim. Using sufficient cribbing to support the tractor is an important safety measure in case the lifting system fails. If you lift an inch, you must crib an inch. Leave no room for error.

Use cribbing blocks of hardwood, such as oak or maple, which will resist splitting. (Do not use concrete blocks for cribbing; they may break.) Handle cribbing by the ends as you place it under the lifted tractor. When placing cribbing, do not put your arm under the tractor. Once cribbing has been positioned, secure it with blocks or chains to prevent slipping.

Lifting a Tractor

1. **Caution:** It is critical that the tractor not slip sideways or shift when lifted. Avoid lifting the tractor any higher than is necessary to extricate the victim. Have nonessential rescuers stand back at least the length of any lifting cables or chains to prevent serious injury if cables should break. Lifting is preferred to rolling the tractor from the victim. When a tractor is rolled by raising one side, the other side may sink further into the ground and injure the victim even more. If the tractor must be rolled, careful cribbing will reduce settling of the lower side.

2. You can use high-pressure air bags effectively to lift a tractor (figure 2.12). Two bags can be stacked on top of cribbing for greater height when necessary. It is important for rescuers to be familiar with the weights of agricultural tractors used in the community so that they can equip themselves with appropriately sized air bags. Rescue units that do not have adequate air bag equipment should at least make cooperative arrangements with neighboring rescue units that can lend additional lifting resources.

   Caution: Tractor weights can be deceptive. Most large farm tractors weigh at least 10,000–20,000 pounds. The largest models weigh more than 40,000 pounds. Be sure to use equipment designed to lift the weight of the tractor involved.

3. Rescue hydraulic lifting units and mechanical jacks of a 5-ton capacity and up (figure 2.13) can be used to lift most tractors, because only a part of the weight must be lifted. Two to three feet of solid cribbing will likely be required to allow the jack to reach an axle or other solid part of the frame. Crib the axle on both sides to prevent the tractor from rocking onto the victim.

4. As an absolute last resort, lift the tractor with another tractor or a wrecker. **Caution:** This method offers the least control and may cause serious injury to the victim or rescuers if the tractor unexpectedly shifts or falls or if the lifting system fails.

Figure 2.12
Air bags

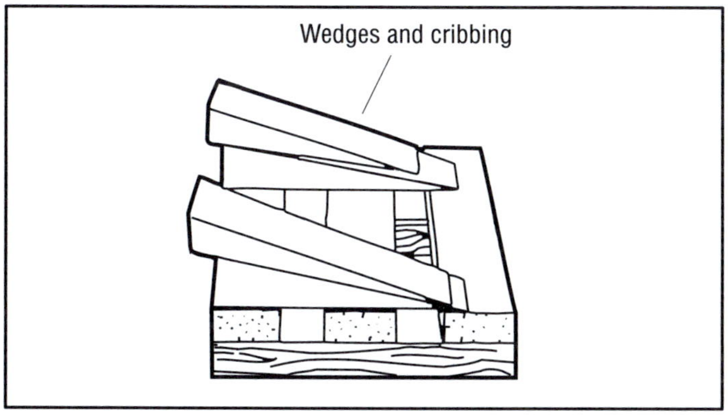

Figure 2.11
Wedges and cribbing

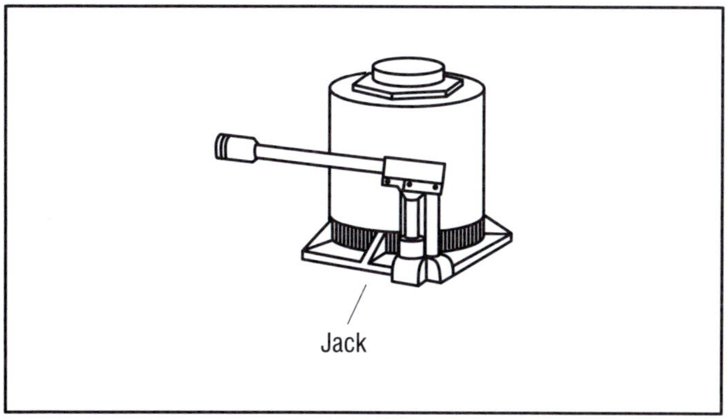

Figure 2.13
Hydraulic jack

Real-Life Scenario

While mowing grass along the steep bank of a manure lagoon, a farmer lost control of his tractor, and it overturned sideways. The tractor landed upside down, pinning him under the seat. From a distance, a worker saw what had happened and rushed to the barn to telephone for assistance.

Within minutes a fire rescue unit arrived at the scene. It was followed shortly by an emergency medical unit. After assessing the situation, they made a radio call for a large wrecker. Meanwhile, as a precaution, a charged fire hose was laid to the site, and CO_2 fire extinguishers were placed in accessible locations.

Before the wrecker arrived, rescuers made an attempt to dig underneath and remove the victim. However, the steepness of the terrain and instability of the tractor made this approach hazardous for rescue workers, so they decided that the rescue could not proceed without a wrecker. While they waited, it was difficult to administer emergency medical treatment because of the victim's position beneath the tractor.

When the wrecker arrived, rescuers carefully attached its power boom winch to the tractor. Then they slowly raised and cribbed the tractor until the victim could be removed.

As is the case in many tractor overturns, the farmer died shortly afterward as a result of a crushed chest and massive head injuries.

Power Takeoff Driveline (PTO) Entanglements

The power takeoff driveline (PTO) transfers power from the tractor to trailing equipment or stationary implements (figure 2.14). All farm tractors are designed to operate the PTO driveline at either 540 or 1,000 revolutions per minute. Universal standards specify that the PTO must operate in a clockwise direction as the tractor PTO stub shaft is viewed from the rear of the tractor.

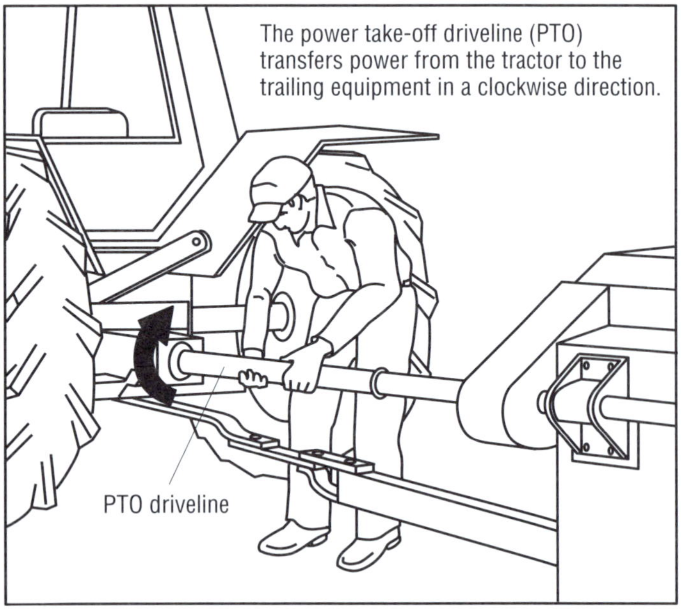

Figure 2.14
Connecting the power takeoff driveline (PTO) for power transfer

The PTO driveline usually consists of a solid shaft that fits into a hollow sleeve and is free to "telescope," or slide together (figure 2.15). A slip clutch and/or shear pin at the shaft end nearest the PTO-powered implement prevents damage to driveline components from overloading. Universal joints or yokes connect the shaft ends to the tractor and the implement. The most common types of PTO couplers are spring-loaded pushpins (figure 2.16) and sliding collars (figure 2.17). Most entanglements involve spring-loaded pushpin couplers. The PTO driveline on most modern farm equipment is well protected with safety shields. Shields on older equipment, however, may not provide full protection, or shields may have been damaged or removed for service or maintenance and not replaced.

Injuries

An unprotected revolving PTO driveline that catches a victim's clothing or hair can rapidly entangle him or her, with catastrophic results (figure 2.18, page 14). Many PTO-related entanglements result in death. They may also involve amputations, severe lacerations, and multiple fractures. Extrication becomes necessary when a victim's limb(s) or clothing has become tightly wrapped around the

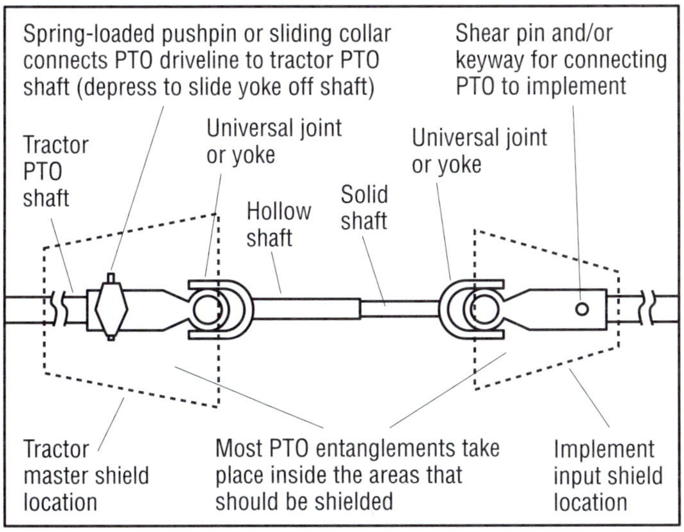

Figure 2.15
Power takeoff shaft (PTO) components

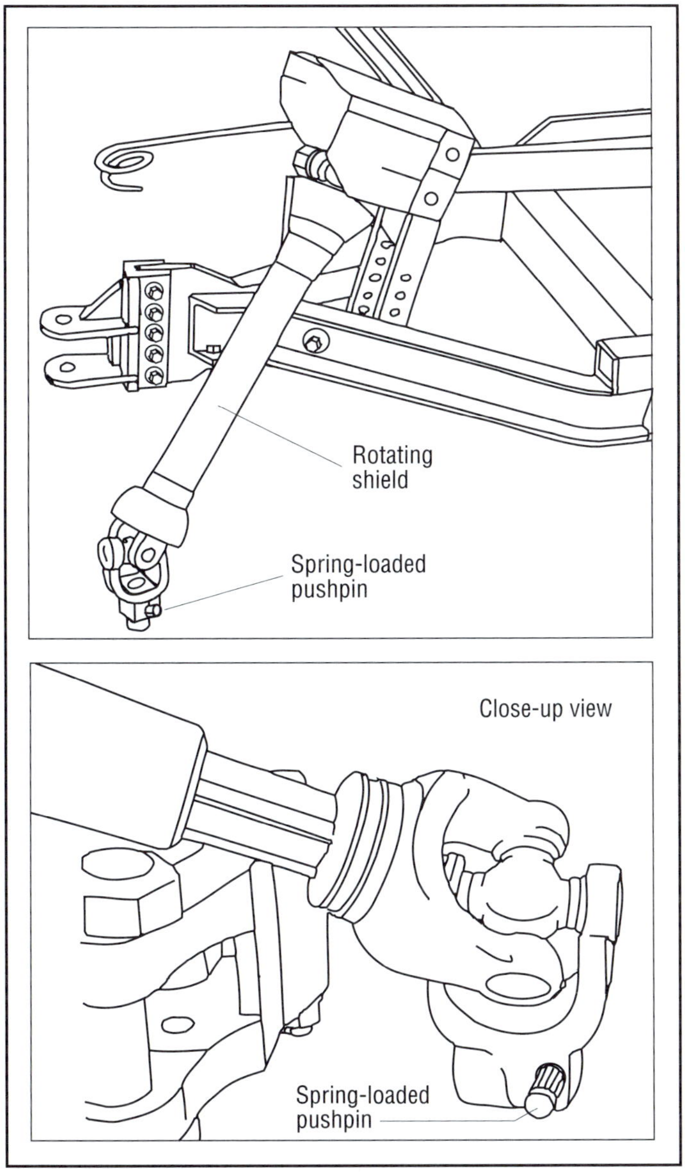

Figure 2.16
PTO coupler with a spring-loaded pushpin

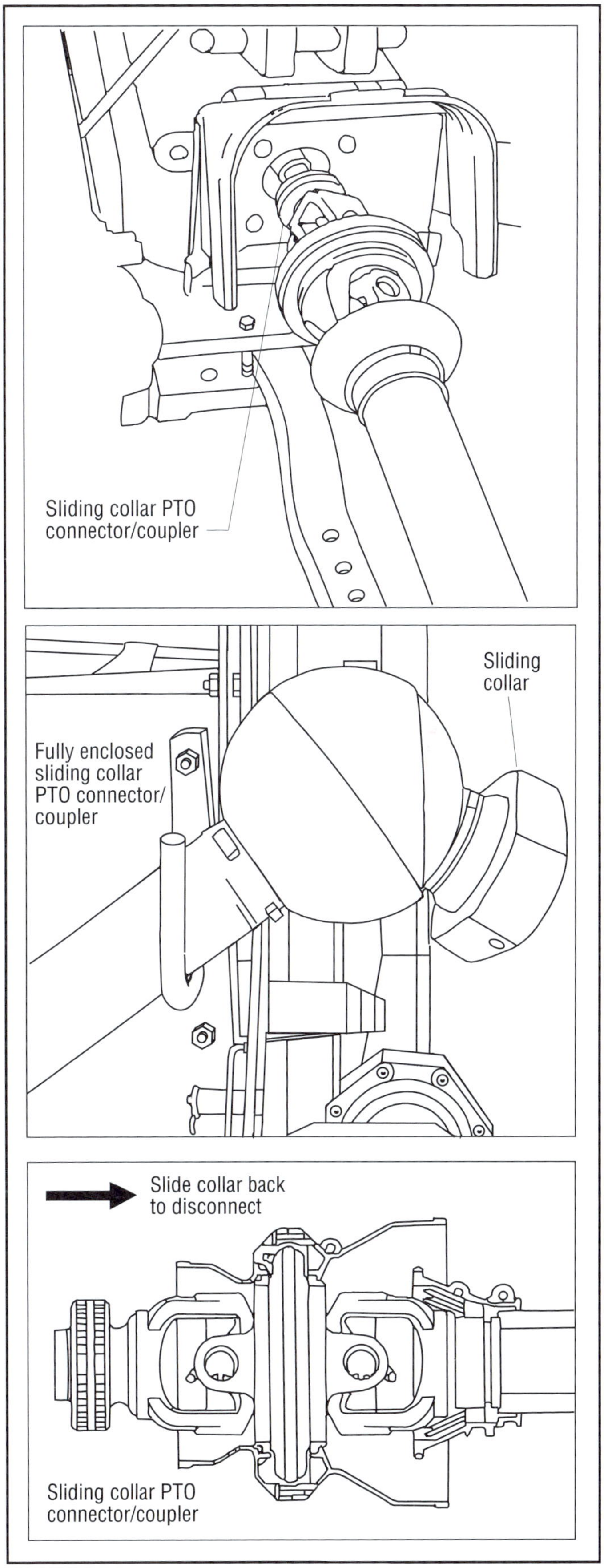

Figure 2.17
Sliding collar PTO connector/couplers
Portions adapted from Bondioli and Pavesi product literature

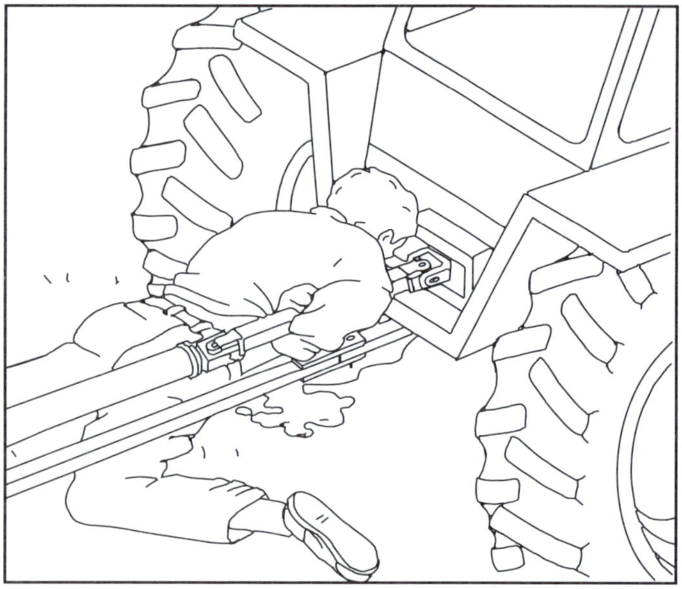

Figure 2.18
This victim, who became entangled around the PTO driveline, is suffering from severe bleeding and arm injuries that may result in amputation.

driveline. Because heavy bleeding is likely, the patient should be monitored for signs of hypovolemic shock throughout the rescue.

Rescue Procedures — PTO Driveline Entanglements

Essential Safety Precautions

1. Spine and neck injuries are common in PTO entanglements. Follow appropriate emergency medical procedures; and stabilize the scene, machinery, and patient.

2. Crib the implement so it is firmly supported throughout the rescue, and block the tractor wheels so that it will not roll. Be sure the tractor cannot start.

   Caution: On some older models, turning the PTO driveline could start the engine.

3. Do **not** allow searching for amputated tissue to delay transporting the patient to the hospital. Any tissue found later can be sent to the hospital separately. (See "Preserving Amputated Tissue," page 4.) Notify the hospital in advance that you are transporting amputated tissue.

4. In some cases, it may be better to transport the stabilized victim still entangled with the driveline. A surgeon can better complete extrication at the hospital.

Extrication Procedures

1. Block the equipment so that it is firmly supported. Be sure the tractor cannot start.

   Secure the PTO driveline at both ends so that it cannot unexpectedly rotate during the extrication process. A pry bar passed through one of the universals and then secured is one approach (see figure 2.15, page 13, for components).

2. If you are unfamiliar with the equipment, don't spend a lot of time trying to figure out how it works. Before attempting to free the victim, seek advice about disassembly procedures from a nearby farm machinery dealer or farmer who is familiar with the equipment. Finding out more about the entangling equipment is likely to be more efficient—and safer—than the trial-and-error method.

3. In the case of entanglement in three-point-hitch-connected PTO-driven equipment, the victim will be less accessible and the rescue more difficult due to the limited space. Disassembly of the three-point-hitch arms that connect the tractor to the implement may be necessary.

Separating the Victim from the Driveline

Telescoping the Driveline

One approach to separating the victim from the driveline that avoids unwrapping him or her and risking additional injuries is to disconnect the driveline from the tractor or implement and telescope the driveline apart. This leaves the patient entangled on a portion of the driveline for transport to the hospital.

Caution: Proceed slowly, and be certain that both ends of the driveline are supported and that the victim is well supported against the shifting or pulling of driveline parts during this process. Before the yoke will slide off the tractor stub shaft, you must depress the front yoke locking pin firmly or slide back the locking collar. Sometimes the PTO driveline cannot be telescoped together enough to remove the yoke from the tractor stub shaft. In such cases, after removing the hitch pin and securing the implement, try to manually roll the tractor ahead slowly (do not start the tractor), either to pull the PTO driveline apart or to slide the stub shaft out of the front of the yoke. (Often the yoke is difficult to remove from the tractor stub shaft. You may need a pry bar to force the yoke back).

Rotating the Driveline

Sometimes gently turning the driveline counterclockwise can help loosen an entangled victim. (Face the rear of the tractor to determine which direction is counterclockwise.) This approach, however, may cause additional injury. In other cases, neither the tractor PTO drive assembly nor the implement drive is designed to allow the PTO driveline to be turned. In extrications involving such equipment, consider disassembling the slip clutch or removing the shear pin at the implement end to allow one end of the driveline to rotate. Once the driveline is free to turn, use a large pipe wrench or a pry bar slipped through the yoke (not on the external rotating shield) to slowly rotate the driveline counterclockwise.

Cutting the Driveline

Some PTO drivelines are one piece and do not telescope. You may have to cut or disassemble this type of shaft at both ends to free the victim. Make sure the implement is stable and secure before cutting.

Air and hydraulic tools that are normally used for shearing metal during automobile extrication have not proven to be effective in cutting solid PTO drivelines. Heavy duty cutters—such as reciprocating saws, abrasive wheels, oxyacetylene torches, or exothermic cutters—have been successful. When using heat-producing tools, consider whether heat will be conducted to the victim and whether there is a danger of igniting combustible material. If available, a heat shield should be used to prevent heat transfer through the metal to the victim.

Before cutting the driveline, disconnect or remove all chains or belts at the implement gearbox to release any tension in the machine. The PTO driveline may be under tension due to stored energy within the machine. If it must be cut, be careful that it does not suddenly snap free and injure a rescuer or further injure the victim.

Minor Entanglements

If only clothing is wrapped around the shaft, you can usually free the victim by cutting the entangled clothing with a razor or sharp knife.

Real-Life Scenario

Two men were using a tractor with a poorly shielded PTO-driven post-hole digger to set posts for a new fence. The auger had difficulty penetrating one area of the field, so one man pressed on the auger gear box to provide additional weight. His coat sleeve caught in the PTO universal joint attached to the gear box. The force of the rotating PTO driveline pulled him into the frame of the digger and threw him around the shaft several times before the tractor engine stalled. Nearly all of the victim's clothing was torn from his body, and he was tightly entangled around the PTO shaft. The other man, shocked by what had occurred, raced to the house to get help.

When they reached the scene, the emergency medical and rescue units secured the equipment and stabilized the victim through appropriate emergency medical procedures. The entangled clothing was cut free, which revealed that the victim's arm had been nearly severed. The PTO shaft was disconnected from the tractor to allow the rescue workers to swing the driveline to one side and free the victim from the frame of the digger. The PTO driveline was separated to free the entangled arm by sliding the tractor end of the driveline off the end connected to the digger. Once free from the digger, the victim's injured arm was immobilized, and he was transported to the hospital.

Hydraulically Operated Equipment Failures

Hydraulic systems reduce the need for complex mechanical linkages and allow for remote control of numerous operations. An engine or motor on a machine with a hydraulic system drives a hydraulic pump, which forces oil from a reservoir through hydraulic lines to a motor or hydraulic cylinders. Valves of varying design and complexity control oil flow to operate the mechanism. Hydraulic systems are used to lift implements such as plows, change the position of implement components such as combine headers or bulldozer blades, operate remote hydraulic motors, and assist in steering and braking.

Single-acting hydraulic systems function somewhat like hand-operated hydraulic jacks. Only one hydraulic line goes to the cylinder from the pump. When the control valve is opened, oil pumps through a one-way valve into the cylinder and lifts the equipment. When the control is released, the lifted equipment is held in place by the pressurized oil. When the valve is released, oil returns to the reservoir and either the weight of the equipment or a heavy spring acts to return the cylinder to the rest position. Combine header lifts, older front-end loaders, and dump beds on trucks and wagons often use single-acting cylinders. Even if the pump is not being operated, single-acting cylinders can often be lowered (retracted) by activating the control lever.

Double-acting cylinders can exert hydraulic force in two directions and are used, for example, on high-lift dump wagons, loaders, backhoes, and tillage equipment, which require downward pressure. Two hoses connected to the cylinder supply fluid to either end (figure 2.19). This equipment is hydraulically locked in position when the control valves are closed; it will not release until the engine is started and the hydraulic valve is activated or the hydraulic hoses are disconnected and bled.

In both single- and double-acting hydraulic systems, there is the potential for equipment to lower unintentionally due to line failure or to leakages past worn cylinder seals or valves in the system.

Injuries

Hydraulic systems operate at extremely high pressure—3,000 pounds per square inch (PSI) or more—and can support tremendous loads; failure of these systems can cause several types of serious injury. Loss of hydraulic pressure can cause supported implements or components to collapse and crush or entrap an individual. A ruptured hose can spray and burn a victim with hot hydraulic fluid. Even pinhole leaks can cause fluid to penetrate tissue and inflict injury on the eyes and body. Hydraulic motors and pumps can also burst and burn the victim.

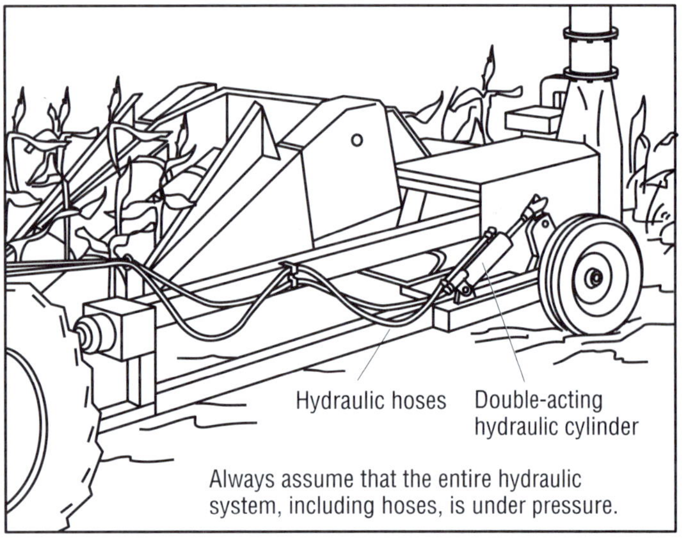

Figure 2.19
The hydraulic system

Caution: If any hydraulic fluid is injected into the skin or under a fingernail, the danger of gangrene exists, which may result in the need for amputation. To prevent dangerous infection, it is essential that the fluid be surgically removed within a few hours by a doctor familiar with this type of injury. Ask the attending physician to call the following toll-free telephone number: 1-800-822-8262. This phone line (operated by John Deere and Company) is open twenty-four hours per day to provide information on treating injuries involving hydraulic fluid. Experience has indicated that many doctors are not aware of the best treatment for such a condition.

Rescue Procedures — Hydraulically Operated Equipment Failures

Essential Safety Precautions

1. If a truck or trailer bed has collapsed on a victim, you may have to remove part of the load before portable jacks can safely lift the bed.

2. Always assume that the entire hydraulic system, including hoses, is under pressure. The pressure of trapped fluid in hoses and cylinders, especially on hot days, can exceed 3,000 pounds per square inch (PSI).

3. Hydraulic fluid is flammable. Avoid open flames and sparks if spilled hydraulic fluid is present. Have a charged fire hose (1½-inch minimum) or an ABC-type fire extinguisher close at hand during extrication.

Releasing Hydraulic Pressure

1. After securing the equipment, it may be necessary in some cases to release the hydraulic pressure in the system in order to move machinery components so that the victim can be freed.

2. Secure or shore up the equipment before attempting to release the pressure. Once pressure is released, the total hydraulic system may become inoperable and parts held by hydraulic pressure may collapse or change position (figure 2.20). In some cases, you may be deal-

ing with a hydraulic system in a series, which could result in more than one machine operation occurring when the pressure is released. In a system of this type, releasing the pressure at one point on a machine may cause unexpected movement in another part. Secure all components in the transport or raised position. Some equipment has built-in jacks, locks, or pins for this purpose.

3. If the pressure cannot be released by the controls, try to disconnect the hydraulic hose at the tractor (figure 2.21). Most hydraulic couplings disconnect by pushing or pulling the sliding collar on the coupling (figure 2.21) The couplings are similar to those found on compressed air lines. Newer, quick-disconnect couplings can be removed by simply pulling the coupling from the valve assembly (figure 2.22).

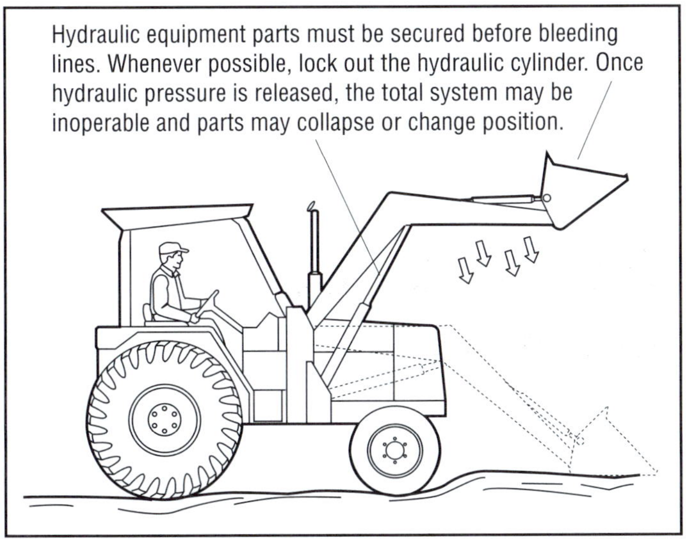

Figure 2.20
Securing hydraulic equipment

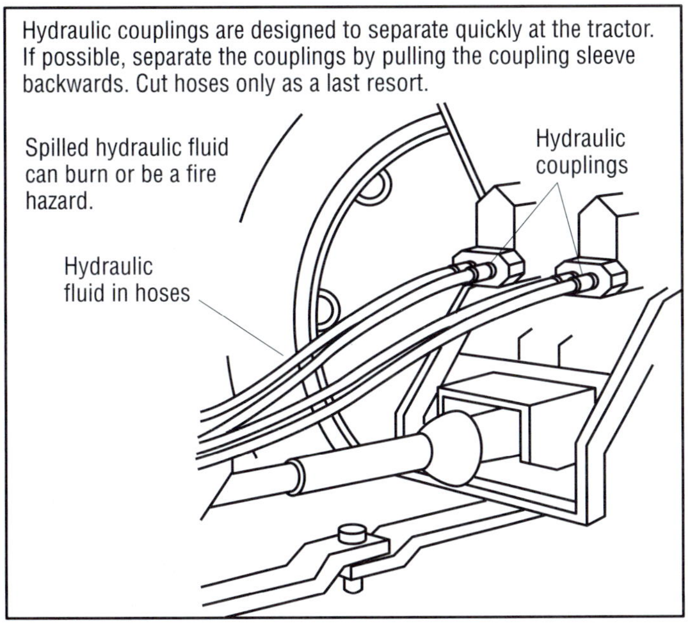

Figure 2.21
Disconnecting hydraulic couplings

4. If couplings are inaccessible, bleed the hydraulic line by opening connections at the hydraulic cylinder with two wrenches. Use full turnout gear to prevent injury to rescue personnel by high-pressure hydraulic fluid. Cover the connections with rags to contain any oil spray.

5. Be careful **not** to cut hydraulic hoses. Hot fluid will spill and create a potential fire hazard. If under pressure, the fluid will spray over a wide area and possibly contaminate open injuries or cause burns. *Note: Hydraulic hoses are constructed of multiple layers of steel webbing covered with rubber and cannot be cut in two with a knife.*

Freeing the Victim

1. Stabilize the scene, shut off the engine, and stabilize the equipment.

2. Always lift and crib the machines manually to remove

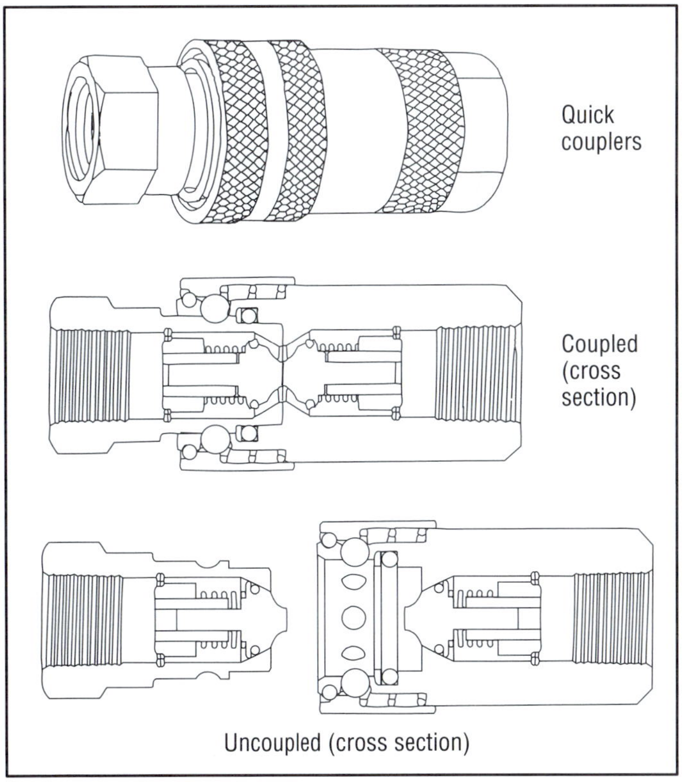

Figure 2.22
Quick-disconnect hydraulic couplings
Source: Fundamentals of Service—Hydraulics, *John Deere Publishing*

the victim. Never try to use the machine's own hydraulic system to lift components off the victim. Assume a failure in the system caused the entrapment.

3. Sometimes it is easier and safer to dig the victim out when you cannot lift the equipment. Crib the equipment to be sure it cannot fall back on the victim.

4. High-pressure air bags have been used successfully in cases where hydraulically operated components, including combine headers and front-end loaders, have collapsed on individuals.

Real-Life Scenario

While a farmer was working under the corn header of his combine, the header slowly lowered over him due to a slow hydraulic leak. The farmer was entrapped beneath the header, which was still partially supported by the hydraulic system. He was discovered some time later by his wife, who called for emergency assistance.

Using standard hydraulic jacks and wooden cribbing, rescue personnel and neighboring farmers were able to raise the header and extricate the victim. With the cribbing, rescuers successfully overcame the problems of the soft ground and the lack of a horizontal underside on the header under which to place jacks.

It was unlikely that the victim would have survived if the residual pressure in the hydraulic system had not kept the full weight of the header from resting on him.

Mixer Wagon and Self-Unloading Wagon Entanglements

Self-unloading wagons (figure 2.23) and mixer wagons (figure 2.24) transport and quickly unload grain, silage, chopped forage, stover, and other loose material. Operated by the PTO driveline from the tractor or by remote-controlled hydraulic motors, the wagon components run with considerable speed and force. The most frequent injuries involving these wagons result from entanglements in the PTO driveline or belts, chains, gears, and other driveline components that are left unguarded when shielding is not replaced.

Entanglements in beaters and augers, however, present the greatest threat of fatal injuries. In some cases, victims have been completely dismembered because of the aggressive action of the augers and rotating beaters used to convey and break up the material being handled.

Injuries

Injuries associated with mixer wagons and self-unloading wagons are diverse. They range from finger and hand injuries caused by entanglement in belt and chain drives to dismemberment resulting from entanglement in PTO drivelines, beaters, and mixing augers. The most frequently reported injuries involve the hand or arm.

Rescue Procedures — Mixer Wagon or Self-Unloading Wagon

Essential Safety Precautions

1. Shut down any running equipment. An elevator, portable auger, silage blower, or tractor is often used with self-unloading and mixer wagons. Turn the equipment off and remove the key.

2. If the entanglement involves the PTO driveline, use the steps described in the section on PTO driveline extrication (see page 14).

Freeing the Victim

1. If the victim is only minimally caught or pinched in the large mixing auger or augers inside the wagon, first make sure there is no stored energy, and then disconnect the PTO driveline from the tractor. After the PTO is disconnected from the tractor, use a large pipe wrench or pry bar to slowly turn the PTO clockwise. Face the wagon to determine which direction is clockwise (figure 2.25). Often you can disconnect the PTO driveline

from the tractor and turn it at right angles to use as a lever to reverse the mixing auger mechanism. Never use power from the tractor to free the victim. If you cannot turn the driveline, trace the drivetrain. Look for hinges and latches that allow housing or guards to be opened to give access to the drive assembly.

Caution: Do **not** attempt to manually rotate the auger in cases where the patient is physically entangled or wrapped on the auger. The sharp edges of the flighting (exposed screw) and the rotating motion of the auger may cause further injury.

2. If the patient is deeply entangled in the auger assembly or if the auger cannot be reversed, disassemble the auger from the wagon and lift the patient and auger out. Cut the drive chain or belt away and remove the bearing housing on each end of the auger. Support both ends of all remaining components during the disassembly to avoid unexpected movement. A heavy plank laid across the top of the wagon and a cable come-along could be used to lift the auger.

   Nearly all rubber "V" or drive belts for agricultural machinery contain numerous strands of tough steel wire and cannot be cut with a knife. In situations where the chain is too heavy for available cutters, use an abrasive cutoff saw; or locate the master link (similar to those found on a bicycle chain) and separate it (figure 2.26). Usually a pair of pliers or a screwdriver is sufficient to separate the master link. This may be difficult to do, however, if the chain is under tension.

3. Consider dismantling the auger housing at the end op-

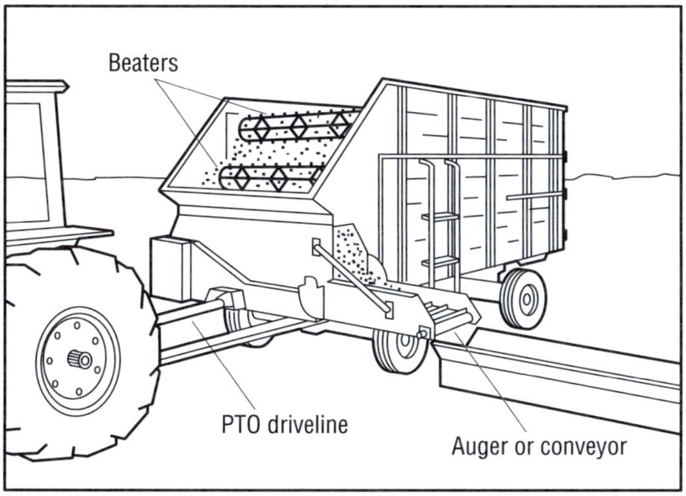

Figure 2.23
Self-unloading silage/forage wagon

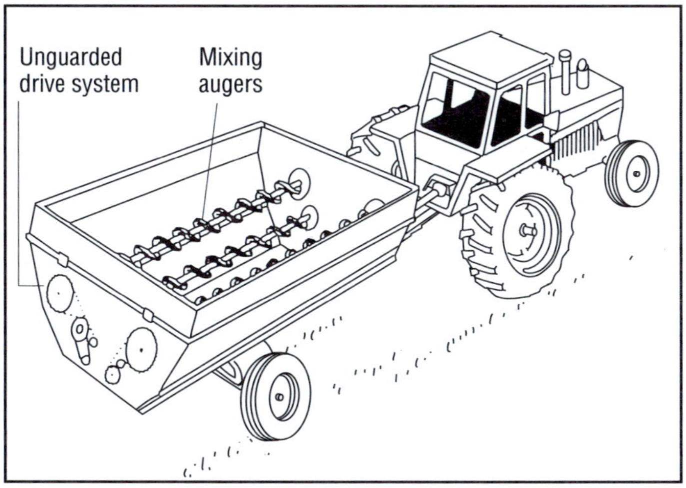

Figure 2.24
Mixer wagon

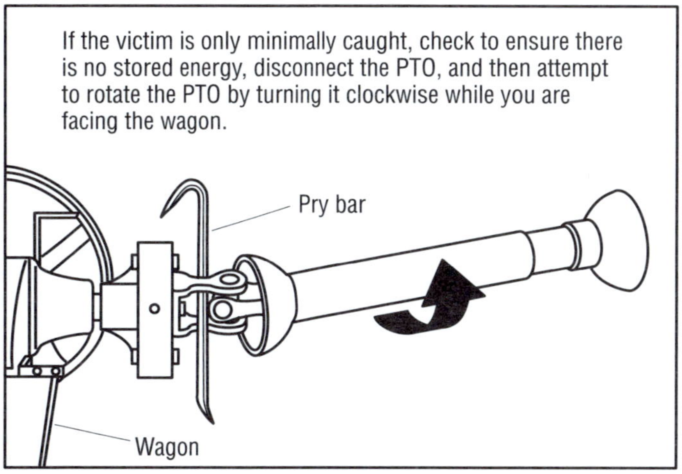

Figure 2.25
Rotating the power takeoff driveline (PTO). Before disconnecting the PTO, make sure there is no stored energy.

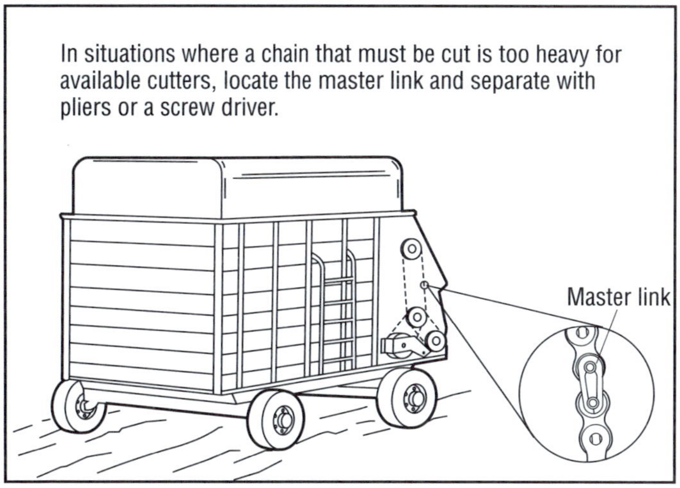

Figure 2.26
Locating and separating the master link

posite the power-drive connection. If you can free this end, either by hand or using a hoist, try to lift the auger enough to free the victim. The end of the auger shaft inside the wagon may have to be cut to allow the auger to be lifted past the end wall of the wagon. Proceed slowly and cautiously so that parts of the equipment do not collapse.

4. In cases of beater entanglement on unloader wagons, the same approach can be used. Attempt to cut or disassemble both ends of the beater and lift it free. Ladders or platforms will be needed to reach the beaters. Beaters are likely to weigh 100–150 pounds; and an overhead lift of some sort, such as a front-end loader, may also be necessary. Use an air chisel or abrasive cutoff saw if possible.

   A somewhat riskier approach is to use an oxyacetylene torch or exothermic cutter to cut one or both ends free (figure 2.27). If an oxyacetylene torch or exothermic cutter is used, guard against fires and place a heat shield between the location of the cutting and the victim to prevent burning him or her.

5. In some instances, it may be appropriate to transport both the patient and the entangling component to the hospital. This is especially true if the victim has been impaled on the beaters or has a limb tightly wrapped around the part.

Real-Life Scenario

A farmer climbed into the back of a self-unloading wagon to break up a bridge of material behind the beaters. While pushing on the pile with a silage fork, he slipped and fell directly into the revolving beaters. The aggressive action of the beaters forced him between them, causing numerous puncture wounds. The drive mechanism of the beater assembly failed, leaving him impaled on several of the beater tines.

It took rescue personnel over an hour to cut out the top beater assembly and remove the victim off the bottom row of tines. Rescuers considered cutting off each of the tines on which he was impaled, but it was determined that the additional time involved would be too risky considering the amount of blood loss that had already occurred.

Many of the rescuers were surprised that the man survived the incident. The help on the scene of paramedics with advanced life-support training was considered a deciding factor.

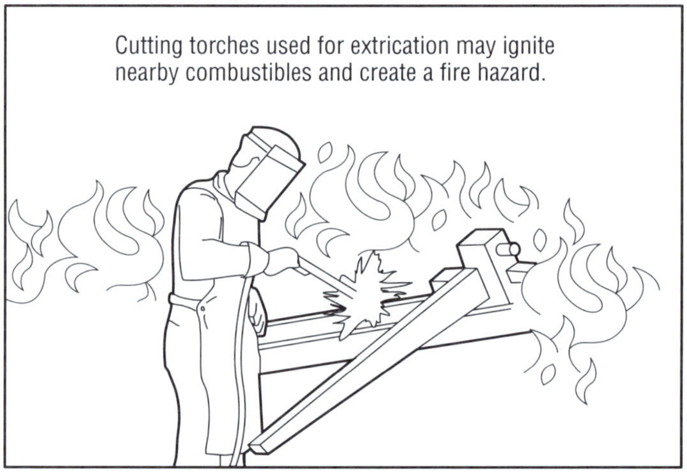

Figure 2.27
Use cutting torches carefully during extrication

Auger and Elevator Entanglements

Portable farm augers transport large quantities of grain, feed, or fertilizer quickly and easily (figure 2.28). Per hour of use, however, the auger is one of the most dangerous pieces of farm equipment. Auger injuries usually involve contact with or entanglement in the exposed screw (flighting) at the intake; entanglement in a belt or chain drive; being struck by an uncontrolled spinning winch crank, which is used to raise or lower the auger (figure 2.29); auger collapse; entanglement in the PTO driveline; or electrical injury when the auger contacts overhead power lines. With the exception of contact with or entanglement in the auger, the same injuries are also common to farm elevators (figure 2.30).

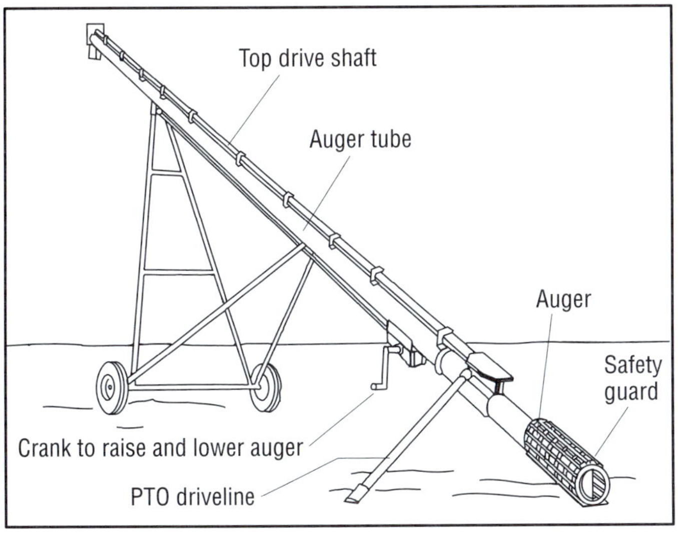

Figure 2.28
Portable auger

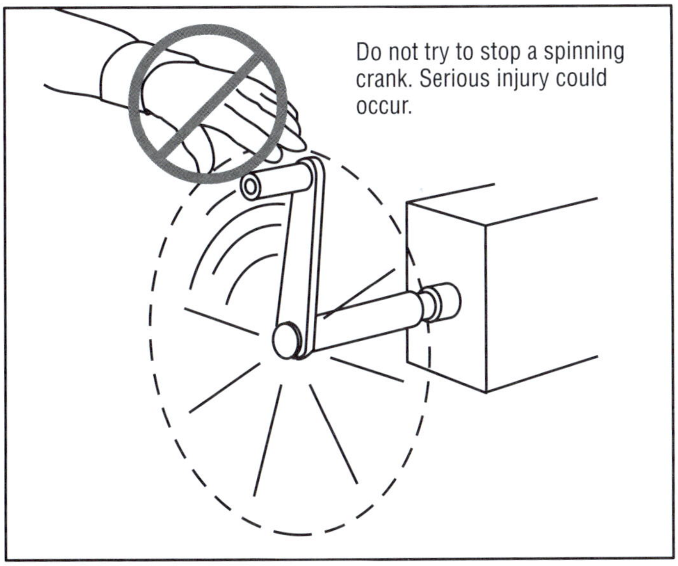

Figure 2.29
Spinning winch crank for raising and lowering the auger

Figure 2.30
Farm elevator
Adapted from New Holland product literature

Injuries

Lacerations, fractures, and amputations are typical injuries caused by augers. The high-speed cutting action, the shear point formed between the auger flighting and tube, and the aggressive action of the auger can result in almost instantaneous amputations. In fact, several studies indicate that portable augers are a major cause of farm-related amputations. Severe lacerations and amputations may cause considerable blood loss. In addition, cases have been reported where the victim's limb was drawn deeply into the auger, resulting in an extremely complex extrication situation. Other injuries include fractures from spinning winch cranks and crushing injuries caused by an auger or elevator collapsing on the operator. Electrical burns can result from an auger or elevator coming in contact with overhead power lines.

Rescue Procedures — Auger or Elevator Entanglements

Essential Safety Precautions

1. If the auger or elevator is in the raised position, check the area for overhead power lines. If there is any possibility of contact with power lines during the rescue, contact the utility company and request that they come and disconnect the power.

2. If possible, lower the unit to increase stability. This may not be possible if movement would further injure the victim or if the upper end of the auger or elevator is already resting on a building or structure. Use caution when moving the unit to prevent it from teeter-tottering (figure 2.31, page 22).

   Caution: An auger loaded with grain may weigh several thousand pounds, so lower it carefully. If operated incorrectly, the lifting mechanism could disengage and cause the auger to collapse suddenly. In some cases, lowering the auger directly onto the roof of a grain bin can provide the stability needed to carry out the extrication.

3. Secure the unit from additional movement by driving a metal stake or bar through the tongue into the ground or by blocking the wheels.

4. Secure all equipment involved in the incident to guard against unintended starting. Shut off the tractor (and take the key); lock out (figure 2.32, page 22) or tape and guard electric motor switches; and remove the

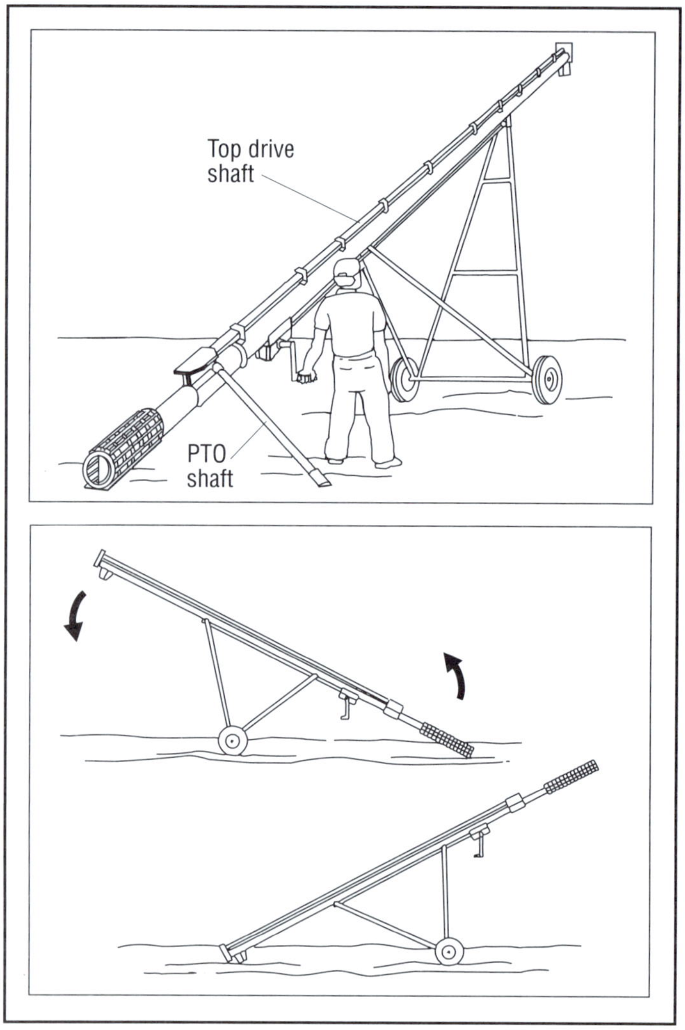

Figure 2.31
Lower the auger to increase stability (top), but avoid a position in which it may teeter-totter (bottom).

spark plug wire from small gasoline engines. Block the auger wheels and tractor wheels to prevent movement. Disconnect the PTO.

Freeing the Victim

Caution: Extrication of a patient severely entangled in an auger is best completed under the direct supervision of a physician at the scene.

1. If the auger is full of grain, an opening should be cut in the bottom of the tube, approximately 2 feet above the entangled limb. This allows the grain in the tube to flow out and prevents it from interfering with the rescue.

2. If the victim's limb is entangled but not amputated, cut the auger tube from around the auger screw and

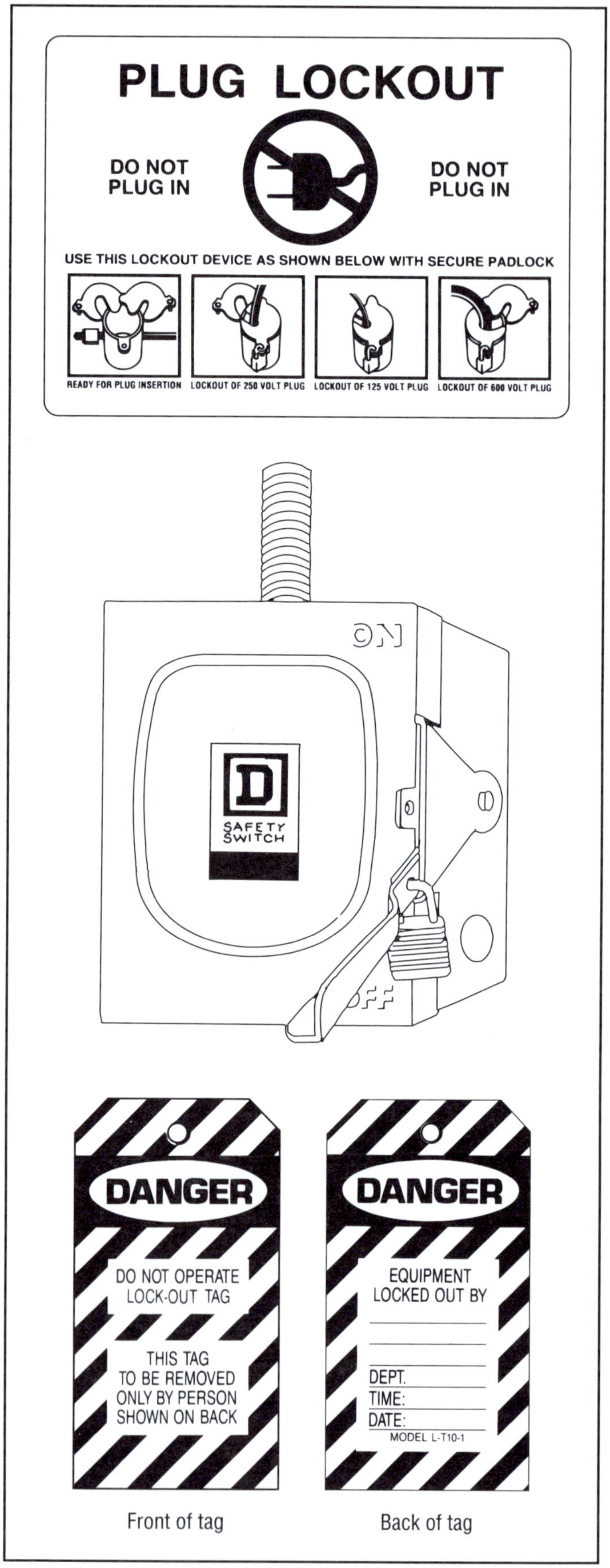

Figure 2.32
Locking out an electrical switch or plug
Portions adapted from material available from Gempler's

limb. **Caution:** Do not reverse the auger because doing so will likely amputate the limb or increase the extent of the injury.

3. One way to remove an entangled limb is to cut 2–4 feet up each side of the auger tube with an air chisel or portable cutting tool that uses an abrasive wheel and operates similar to a portable power saw. Then cut across the top to expose the auger screw and limb. **Caution:** Be sure to protect the victim from flying metal and sparks. Do **not** use an oxyacetylene torch. Care should be taken to prevent the auger from bending and collapsing after the section of the housing has been removed. On large-diameter augers, the bottom portion may be a section of thick-walled pipe that may be difficult to cut.

4. Because an entangled limb may be greatly elongated into the auger, be sure to locate the entire limb and avoid further injuring it when cutting the tube. Exposing the screw and limb will reduce the danger of added injury during extrication. Cut the auger screw above the limb and lift the limb and auger screw from the tube.

5. If the victim is severely entangled, it may be best to support both ends of the auger assembly and cut off 4–6 feet of the inlet end. Generally, the shaft running through the center of the auger screw (or flighting) is hollow and can be easily cut. Once the end is off, you may be able to slide the auger screw and the entangled limb from the tube. If removal is not possible, take the auger section with the patient to the hospital. Hospital personnel can then monitor the patient during auger removal.

6. If the entanglement involves only clothing or does not involve the wrapping of an extremity around the auger, it may be possible to free the patient by reversing the auger. Do this manually by rotating the driveline in the direction opposite to its normal operating rotation (clockwise as you face the gearbox on the auger).

Recovering Amputated Tissue

First check the auger outlet or grain bin for signs of amputated tissue. In some cases amputated tissue may still be inside the auger tube. If the tissue is visible or known to have penetrated a short distance (2–4 feet) into the auger, cut open the auger tubing to remove the tissue. **Caution:** Amputated tissue may be damaged considerably more if the auger is reversed. If the tissue has traveled a considerable distance into the tube, manually rotate the auger in a forward direction to transport the tissue toward the auger's outlet.

Caution: Do **not** use the normal power source to turn the auger. Have someone observe the auger's outlet to prevent the amputated tissue from jamming against the end plate of the auger tube.

Do **not** delay transporting the patient to the hospital while you look for amputated tissue. Quickly transport any tissue found later and notify the hospital that amputated tissue is on the way. See "Preserving Amputated Tissue" on page 4.

Real-Life Scenario

An eleven-year-old boy was helping his father unload grain from a truck into a grain storage bin with an auger. When the truck was empty, the father left the boy to clean up the spilled grain around the auger while he went to check the grain level in the bin. While shoveling grain into the auger, the boy unintentionally threw a cob into the hopper. As he reached in to retrieve it, the auger instantly entangled his arm and completely severed it just above the elbow.

Hearing the boy's scream, the father ran and found him sitting on the ground next to the auger and bleeding heavily. The father folded his shirt several times and used it as a compress against the exposed stump. After the bleeding slowed, he carried the boy to the house to summon medical help.

Upon assessing the situation at the scene, one emergency medical technician administered initial medical treatment to the boy. Another accompanied the father to the injury site, where they found the amputated arm in the grain storage bin directly under the auger outlet. Rescuers wrapped the limb in a sterile towel and packed it in ice from the kitchen refrigerator. They transported the boy and his arm to the hospital, where the limb was successfully reattached.

Electrocution

Contact with overhead power lines causes most on-farm electrocutions and serious electrical injuries. Grain augers, hay elevators, aluminum ladders, radio antennas, and irrigation equipment are frequently involved (figure 2.33). Many overhead power line electrocutions involve more than one person working together at the time of contact (a father and son setting up an antenna or two people moving an auger, for example). Very high-voltage transmission lines often cross agricultural land, though they seldom cross the farmstead. In addition to accidents involving equipment in the farmstead, farmers are electrocuted when raised equipment touches these very high-voltage transmission lines during road transport and when entering fields.

Several factors influence the effects of electrical contact on a victim. These include amperage and voltage, duration of contact, resistance of the body and clothing (particularly footwear or gloves), and the path of current flow. All electrical power lines, from 120-volt household lines to 345,000-volt transmission lines, have the potential to kill or cause severe injuries. Most fatalities involve the 120- to 240-volt lines that feed power from the primary transformer.

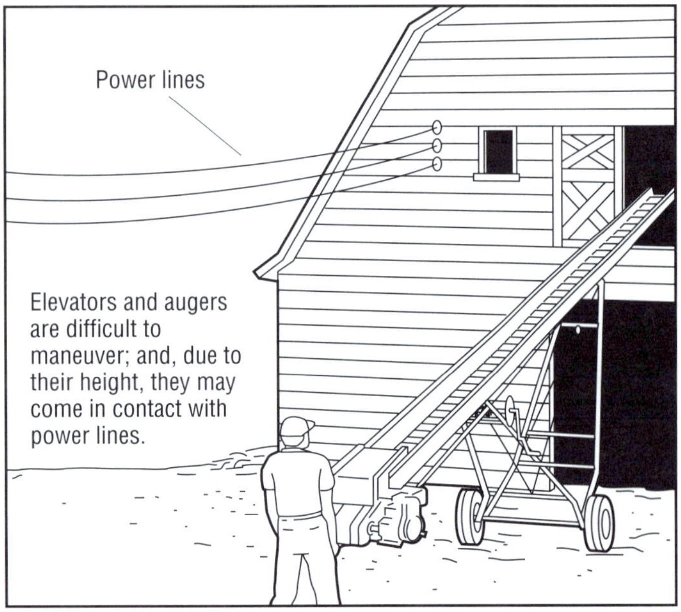

Figure 2.33
Augers and elevators are frequently involved in on-farm electrocutions

Injuries

Electrical injuries are diverse. Severe burns generally result at both the point of contact with the victim and the point at which the current leaves the body to enter the ground. Current flow may also cause internal burns. Clothes may ignite, causing extensive burns to the skin. Injuries to the nervous system are also common. Temporary paralysis may occur, and the victim may also lose some sensory and motor abilities. The full extent of injuries may not be immediately apparent.

Respiratory and cardiac arrest are the primary causes of death following electrocution. Individuals with prior cardiopulmonary irregularities are less likely to survive electrical shock than those without such conditions.

Rescue Procedures — Electrocution

Essential Safety Precautions

1. All overhead or buried power lines should be considered live and insulated, regardless of their appearance. They should never be handled by anyone without the appropriate training and equipment.

2. Always assume that anything—including the victim or machinery—that is in contact with a power line is energized.

3. You don't have to make direct contact with a high-voltage line to receive an electrical shock. Electrical current can reach you through soil, damp wood, and even through the air via arcing.

4. To prevent electrocution of rescuers, notify the power supplier immediately to have a crew respond and disconnect the power source.

5. Fire-fighting turnout gear is *not* designed to protect the user from high-voltage contact.

Rescuing an Electrocution Victim

Caution: The following procedures are not adequate in

incidents involving very high-voltage transmission lines. Very high-voltage transmission lines run along the road or cross country. Immediately contact the power company if a very high-voltage incident occurs.

1. Under no circumstances should a rescue be attempted if the victim or equipment involved is still energized, unless the rescuer is properly trained and equipped. Handling high-voltage lines leaves no room for error.

2. A safe work area should be secured to ensure that no one comes in contact with any energized component.

3. If the contact was with an overhead or buried line that feeds the farmstead, call the utility company first; then protect the site until utility representatives arrive.

4. If the power source is a breaker box, lock out and secure the breaker box after shutting off the main circuit so that no one can unintentionally restore power (figure 2.32, page 22). Do not reenergize the system until the rescue has been completed and the owner or utility representative has inspected the power source involved in the incident.

5. After freeing the patient from electrical contact, check for breathing and initiate basic life support. Bleeding is usually not a problem, though there may be wounds. Treat electrical burns as you would other types of burns.

6. Regardless of how the patient may feel, anyone receiving a high-voltage electric shock should be examined by a physician. The potential for internal injuries exists.

7. Farmers may be electrocuted when raised equipment touches very high-voltage transmission lines during road transport and when entering fields. (In most cases, however, these lines are not found in the farmyard where augers and elevators are used.) The above procedures are not adequate in incidents involving very high-voltage transmission lines. Immediately contact the power company if a very high-voltage incident occurs.

Real-Life Scenario

On Thanksgiving morning, a family had gathered to celebrate. A farmer and his son, who was home for the holiday, headed out early to complete chores, including filling the drier with newly harvested corn. In the process, they moved a portable auger from a full bin to an empty one. One of the two maneuvered the tongue of the auger, while the other pushed on one of the wheels. Without noticing, they pushed the auger into contact with an overhead power line feeding the grain handling center. Both men were electrocuted almost instantly. When other family members came to find them, both bodies were burned beyond recognition and the tires on the auger were still burning.

Harvesting Equipment Injuries and Emergencies

Agricultural operations use a wide variety of harvesting equipment that is designed to cut, chop, grind, compress, and blow crops for transport and storage. This equipment includes balers, combines, cotton pickers, corn pickers, potato diggers, forage harvesters, and forage blowers. Because many of the hazards associated with these machines are similar, rescue techniques are comparable. For the sake of brevity, this chapter focuses on balers, combines, and corn pickers.

Hay Balers

Hay balers compact hay and straw into dense, easy-to-handle bundles. Conventional balers form 40- to 150-pound rectangular bales held together with twine or wire (figure 2.34, page 26). Round balers roll the crop into 100- to 1,500-pound cylindrical bales (figure 2.35, page 26). Large square balers form rectangular bales weighing 1,000–2,000 pounds and bind them together with twine or wire. On the conventional baler, the PTO, the knotting mechanism, bale kicker, and pickup assembly cause most injuries. For small round balers, the knotters and rollers cause the most problems by grabbing and crushing the victim's arm when the victim tries to feed in the twine by hand. In large round balers, the pickup assembly and the belts, chains, or rollers that compress the crop cause most entanglements.

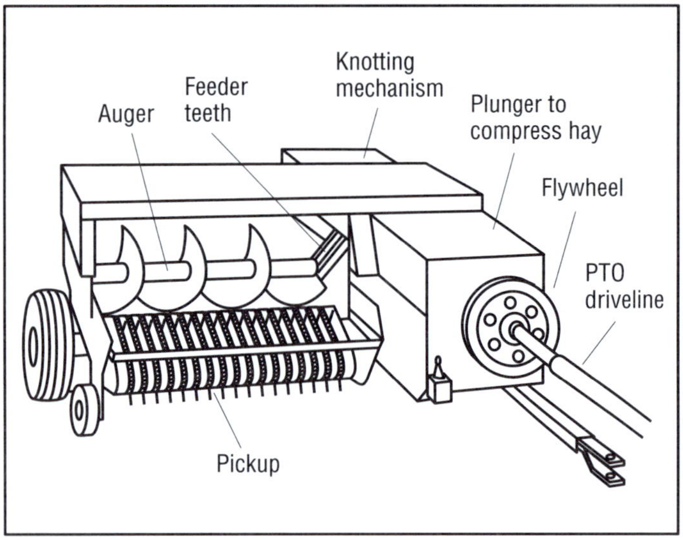

Figure 2.34
Conventional hay baler

Figure 2.35
Round baler

Injuries

Because of the aggressive action of most types of harvesting equipment, there is a high likelihood of fatality when a person becomes entangled in the harvesting mechanism or driveline components. Injuries to the hands and feet, including amputations and crushing injuries, are also typical. A contributing factor to low survival rates in harvesting equipment injuries is the probability that victims in isolated locations will suffer substantial blood loss before being found.

Rescue Procedures — Conventional Balers

Essential Safety Precautions

1. As with all entanglements, shut off the tractor first and remove the key.

2. **Caution:** The risk of fire exists during extrication from a baler. Use a cutting torch or power saw only with utmost care and only as a last resort to extricate the victim. Have a charged fire hose (1½ inches minimum) and fire extinguisher handy. Remove all hay from inside the baler and clear any loose hay surrounding the baler and tractor. Hay and straw stubble are combustible and should be protected from sparks.

Freeing the Patient

1. Never use power from the tractor to attempt extrication. The moving components on most conventional balers can be reversed by hand, after the PTO driveline is disconnected from the tractor. Rotating the large flywheel located at the front of the baler clockwise when viewed facing the front of the baler will reverse the feeding mechanism.

2. If a hand has been entangled in the knotter mechanism, disassembly is usually the best approach. A farm equipment mechanic with baler experience can disassemble a knotter quickly. If a mechanic is not available, close examination of the knotter components should make disassembly procedures apparent. (Check the baler toolbox or twine box for an operator's manual, which is often helpful.) Even the patient, in some circumstances, can provide information on disassembly.

3. In some cases the victim may have become entangled in the cross auger found on the pickup assemblies of some balers. This auger is usually supported on only one end and can easily be removed. Block the pickup to keep it from moving during extrication.

4. Entanglement in the pickup assembly may involve impalement on the pickup tines (sharp projections). A heavy bolt cutter can be used to cut off any tines embedded in the patient so that he or she can be extricated, but the tines should not be removed from the patient's body at the site. To minimize jolting the tines during cutting, hold both sides of the tine with vise grips and cut between the grips. Because the removal of tines from the patient can cause severe hemorrhaging, it should only be performed at the hospital.

5. Complex entanglements often require a physician at the scene to supervise life-support procedures during extrication.

Rescue Procedures — Round Balers

Essential Safety Precautions

1. Remember that the risk of fire is present throughout the rescue. Clear loose hay from the area, and have a fire extinguisher and charged fire hose ready.

2. Disassemble the unit with extreme caution since several powerful springs are used to keep the belts under tension. Sudden release of the stored energy in one of these springs can cause serious injury to rescuers.

Freeing the Patient

1. Shut off the tractor and crib any parts that may shift and cause further injury.

2. Victims entangled in the belts/chains that form the hay into a bale may have been pulled in so tightly that they have difficulty breathing. Cut belts with a sharp knife or saw to relieve pressure. Tearing or cutting out surrounding hay will also reduce pressure on the victim.

3. Rollers used to compress the crop as it passes from the pickup into the bale chamber are spring-loaded. Tension can be released by loosening spring or hydraulic pressure on one of the rollers. Wedge the rollers so that they cannot turn. In many cases, air bags can then be used to spread the rollers sufficiently to extricate a victim caught in the pickup assembly.

4. Entrapment between the rear door and frame can occur because the door can close if the victim is working at the back of the baler with the door raised. To release the victim, the door must be raised again. Since most of the doors are raised by double-acting hydraulic cylinders, the lines to the cylinders must be bled while the door is being pulled open enough to release the victim. Do not operate the hydraulic controls unless the operator is absolutely sure of control functions.

Real-Life Scenario

A teenaged boy was operating a large round baler in an isolated field. Attempting to clear a plugged pickup, he left the machine running and began to kick the loose hay into the compression rolls. One foot became entangled, and then, after he struggled briefly, his second foot was also pulled into the machine.

The large flat belts used to form the bales slowly burned off the boy's shoes and then began to burn off his feet and lower leg. There was very little blood loss due to the nature of the injuries.

Over an hour later, another worker discovered him still caught in the operating machine. The employee shut off the tractor and summoned help. Responding rescue personnel released the compression roll adjustment and removed the compression springs to open the rolls enough to release the boy's legs. The abrasive action of the belts left no amputated tissue that could be reconnected. The boy remained conscious throughout the entire ordeal.

Combine Entanglements

Combines have enabled farmers to harvest large acreages of grain in extremely short periods of time, reducing both exposure to the elements and the hazards associated with hand harvesting or using corn pickers. Modern combines are extremely aggressive; operating at high field speeds, they can extend over 30 feet wide and weigh over ten tons. Because they cut, thresh, transport, and clean the grain in one pass, combines are actually several machines in one, with numerous points of potential injury (figure 2.36).

Injuries

Most combine injuries occur during service or maintenance and involve the victim's extremities. Falls from the operator's platform or other working surfaces on the combine are also a frequent cause of injury. The most serious injuries involve entanglement in the harvesting unit or entrapment underneath the harvesting unit or "header." These injuries are often fatal, due to the aggressive action of the machinery involved and the weight of the harvesting unit. Entanglement in the augers used in the grain tank and in the unloading mechanism is also possible. Straw choppers at the rear of the combine can also cause serious injuries.

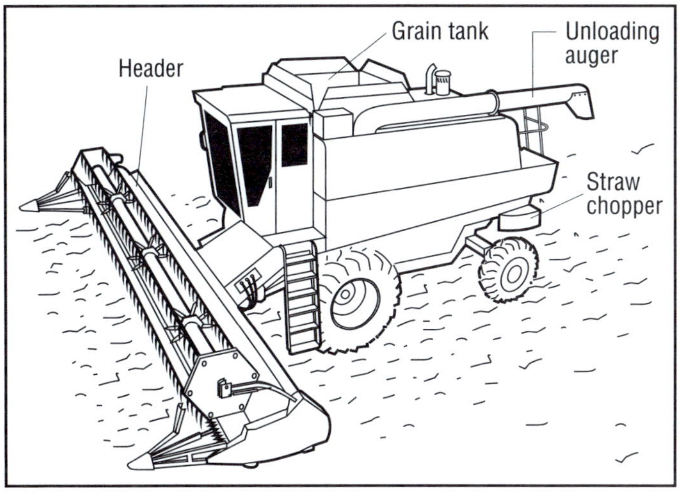

Figure 2.36
Combine with a bean header

Rescue Procedures — Combine Belt or Chain Drive

1. Shut off the engine.

2. Free the victim by cutting the belt or chain with bolt cutters or a hacksaw. In some cases, it may be easier to separate the chain using the master link (figure 2.26, page 19).

3. Do **not** attempt to feed a hand caught in a pulley or gear around the gear or pulley; this may injure the hand worse. Remove the belt or chain from over the top of the hand and pulley or gear.

Rescue Procedures — Combine Grain-Tank Auger

Minimally Caught

You can sometimes release a victim who is caught minimally in a grain-tank auger by reversing the mechanism (turning it in the opposite direction to the flighting's normal operating rotation) by hand or with a pipe wrench. First, shut off the engine and remove the key. On many combines, you must disconnect the drive assembly, usually a belt, before you can reverse the auger. Disconnecting the driveline will allow each auger to be moved separately. Many combines come equipped with a large wrench that can be attached to some point on the drive assembly to reverse the feeding mechanism to free plugged material. This may be useful in some cases of entanglement.

Severely Entangled

1. If the victim is severely entangled, disconnect the auger's driveline first. Do **not** attempt to reverse it manually, or additional injury may occur.

2. Try to dismantle the auger bearing housing at the end opposite the driveline. Once the auger is free, slowly lift it by hand or with a hoist.

3. If you cannot release the auger, try to slowly free the auger shaft with a rescue tool, such as a hydraulically operated spreading tool.

4. If all else fails, cut both ends and carefully hoist the auger free.

Rescue Procedures — Combine Corn Header

1. Due to the aggressive action of the snapping rolls located in the corn header and the massive size and strength of the header, extrication can be difficult.

2. Secure the header, using jacks and cribbing to prevent it from dropping or moving during the rescue.

3. If possible, remove the gathering chain and open the snapping rolls to their maximum opening.

4. Use air bags or hydraulic spreading equipment and wedging to further spread the snapping roll assembly. This assembly is structurally very strong and will require substantial effort to spread.

5. Do **not** attempt to use the power-reversing mechanism found on some newer combines to reverse the snapping rolls, for this would result in additional injury to the patient.

6. In some rare cases it might be necessary for a physician to perform a field amputation to free the patient.

Rescuing a Victim Pinned under the Header

1. Immediately block or secure the header to prevent further lowering due to hydraulic creep caused by hydraulic fluid leakage.

2. Block the wheels so the combine will not roll while the header is being lifted. Be sure to use adequate cribbing, even when only part of the header's weight rests on the victim, as the header may lower unexpectedly.

3. Always lift the header with hydraulic jacks or air bags. If the combine itself raises as the header is raised, the total weight will usually exceed the capacity of most available lifting devices. You can lift the header separately by disconnecting the hydraulic hoses from the hydraulic lift cylinders on the header or by actuating the control valve. **Caution:** Do **not** attempt to disconnect hydraulic cylinders without adequate cribbing if only part of the header's weight is on the patient; releasing the hydraulic pressure will cause the header to lower completely. By using sufficient cribbing, prevent the header from lowering onto the patient.

4. If the ground is soft, it may be possible to dig under and free the victim in some cases. Crib the header to prevent further settling.

Real-Life Scenario

Late one fall a farmer attempted to finish harvesting his last few acres of corn after a light freezing rain. The icy conditions and tough, partially frozen cornstalks had caused severe plugging problems in the combine throughout the day. On one occasion, the farmer tried unsuccessfully several times to clear the snapping rolls on the header with the power off. Finally, in frustration, he started the machine, left the engine idling, and tried again to remove the plug with the snapping rolls turning. The plug suddenly broke loose as the farmer was tugging on it and his arm was pulled into the machine along with the plug. The dense plug and farmer's arm stalled the engine, but the arm was drawn completely through the rolls and tightly wedged between the stripper plates. Almost an hour later the farmer's son returned with an empty grain-transport wagon and found his father badly entangled. Unable to free him, the son raced home to call for assistance.

On the basis of the son's description of the incident, the dispatcher sent both the emergency rescue and medical units to the scene. After surveying the extent of the farmer's injuries and the complexity of the entanglement, the emergency medical technicians called the hospital and requested that a physician come to the scene and be prepared to amputate the arm if necessary. They carried out emergency life-support procedures while waiting for the physician.

Meanwhile, the rescue personnel secured the header with cribbing and jacks. Then they opened the snapping rolls as far as adjustments would allow. Next, they disassembled the driveline to the header so that they could reverse the one set of snapping rolls that entrapped the victim independently of the other rolls.

A physician arrived and decided that rescuers should attempt to reverse the rolls by hand to free the arm. He then treated the victim for possible shock and severe blood loss. The rescue squad used pipe wrenches to slowly reverse the snapping rolls and eventually free the arm. The arm was seriously injured but did not have to be amputated.

The success of the rescue was later attributed to the excellent cooperation among the emergency rescue and medical units, the physician, and the hospital.

Corn Picker Entanglements

Every year farmers are seriously injured when they try to pull a plug of stalks or other trash from running, older corn pickers (figure 2.37) and are pulled into the machines. Entanglement can occur in either the snapping rolls that strip the ears from the stalks (figure 2.38, this page; figure 2.39, page 30) or in the husking rolls that are designed to remove the husks from the ear (figure 2.40, page 30). These in-running rolls form pinch points that can pull a limb in more quickly than an operator can react. Snapping rolls have spiral ribs that mesh together forcefully enough to snap the ear from the stalk (figure 2.39, page 30). Some pickers, such as those found on combine corn heads, have snapping rolls with stripper plates. The rolls pull the ear against a set of plates to snap it from the stalk. Once snapped from the stalk, the ear is carried to a husking bed (figure 2.40, page 30). Pairs of rotating iron and rubber rolls in the bed catch and tear the husk from the ear.

The heavy metal housings in which the snapping and husking rolls are mounted can make extrication of an entangled operator quite difficult. In most cases, it is also very difficult to reverse the rolls. Either the mechanism turns in only

Figure 2.37
Corn picker

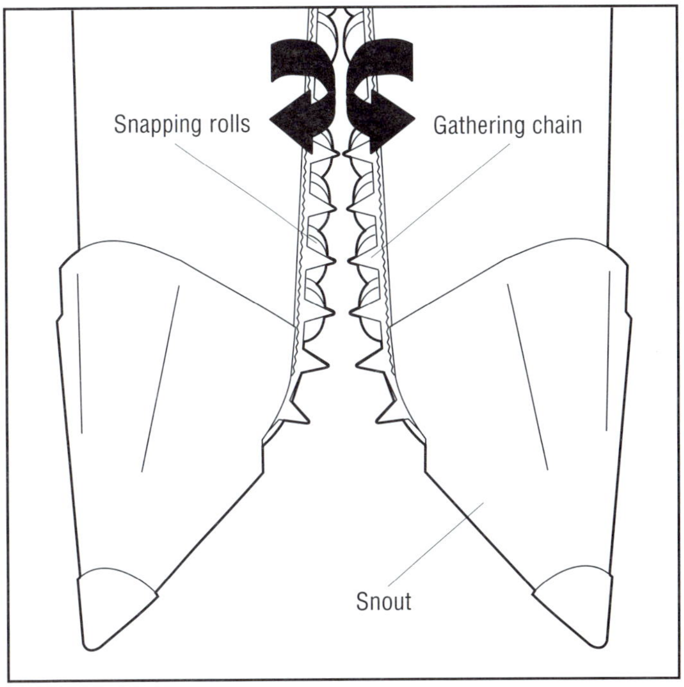

Figure 2.38
Snapping rolls

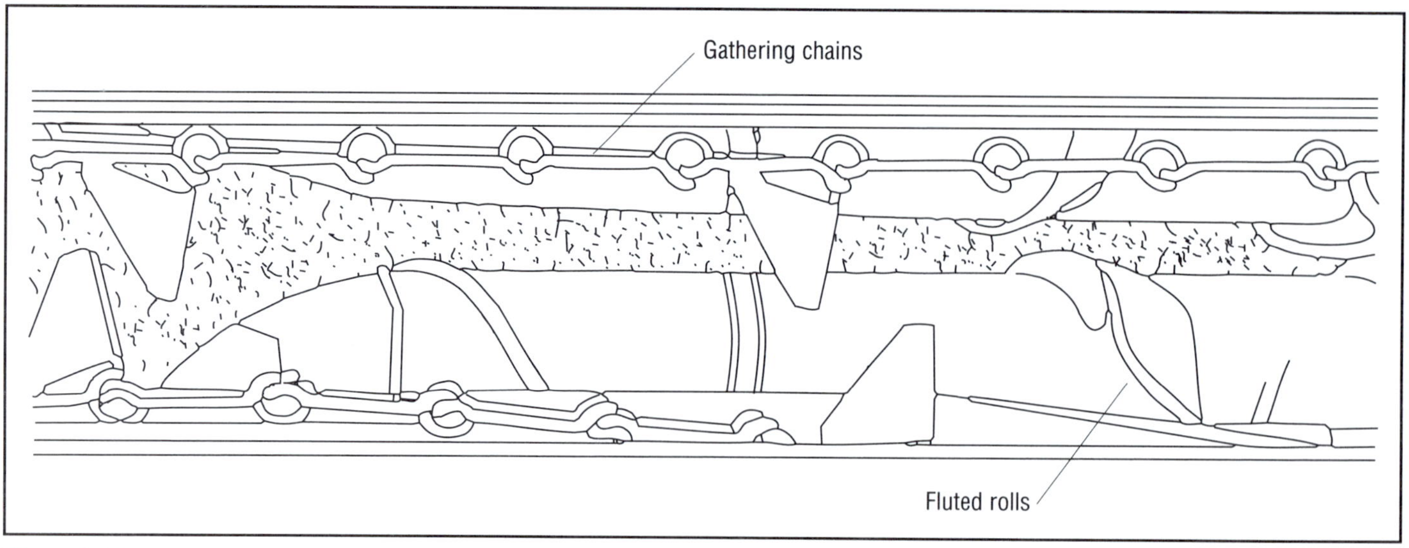

Figure 2.39
Snapping rolls

one direction, or the victim has been pulled too far into the picker for reversing the mechanism to be helpful.

Injuries

Corn picker injuries usually involve the upper and lower extremities; but the most common type of injury is amputation of fingers, hands, and arms. Because of the aggressive action of the snapping and husking rolls, entanglements in these components can result in severe cuts, crushing injuries, and amputations. The in-running rolls can quickly pull one or more limbs deeply into the machine, preventing escape. There have been reports of individuals who have been entrapped in operating pickers for hours before they were discovered. Blood loss and shock are the typical cause of death in cases of picker-related entanglement.

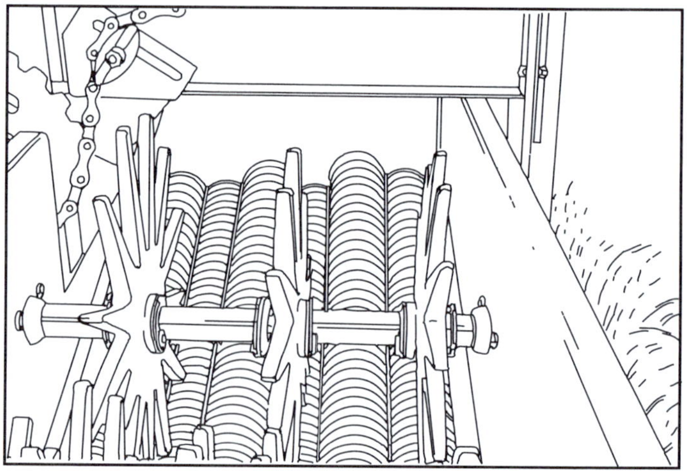

Figure 2.40
Husking bed

Rescue Procedures — Corn Picker

Essential Safety Precautions

1. Shut off the engine. Disconnect the PTO if the picker is tractor-towed or -mounted. Anticipate stored energy when the PTO is disconnected.

2. If no one is present who is familiar with the machine, call a machinery dealer, who can often tell you how to dismantle the picker so as to remove the victim. Older corn pickers may be heavily rusted, making disassembly difficult.

3. Many corn-picker extrications have lasted an hour or more. Since the victim's strength may diminish quickly, a physician should be called to the scene. As with all farm rescues, medical-control communication should be established.

Prying Rolls Apart to Free the Victim

1. On most snapping roll assemblies, the rolls can be adjusted apart a small distance, usually less than 2 inches. It is unlikely that rescuers will be able to adjust the snapping rolls far enough apart to complete the rescue. This opening, however, can provide enough space to get a "bite" with spreading equipment. On husking rolls, typically no adjustments can be made, and some rubber may need to be cut off the rolls to gain a point of access for spreading equipment.

2. Snapping rolls can be pried apart with air bags or a hydraulic rescue tool. Place the prying equipment care-

fully, since it tends to slip on round surfaces. Because air bags will not slip, they provide an advantage for prying apart round surfaces. Do not use cables; they may slip or break without warning.

3. Husking rolls should be disassembled to free the victim because the space between husking rolls does not generally allow for the use of air bags. Hydraulic rescue spreaders with thin tips have been successful when used in conjunction with wedging.

4. Use strong, tapered wooden wedges to hold the snapping rolls apart while the rescue tool is positioned (figure 2.41). Wedges should be used in pairs from opposite directions. Because snapping rolls already have some space between them, they are usually less difficult to pry apart than husking rolls, which are placed tightly together.

5. It is common for the rolls to break or the frame that holds the rolls to bend or deflect during prying.

6. If efforts to pry equipment apart are not successful, a torch, such as an exothermic cutter, may be used to cut the equipment apart. However, this is not a preferred method; cutting may be difficult because many parts are made of cast iron or hard-surfaced steel. When using a torch, have someone stand by with a charged fire hose (minimum 1½ inches) and a fire extinguisher in case combustible materials should catch fire. The patient should also be shielded from sparks and heat that may be conducted from the torch.

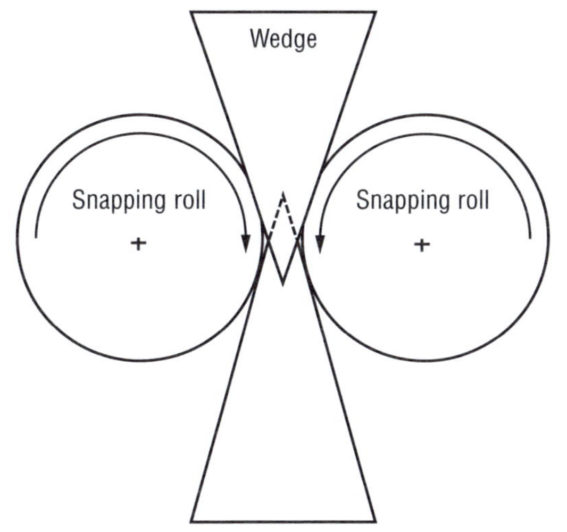

Figure 2.41
Placement of wedges to hold snapping rolls apart and keep them from turning

Real-Life Scenario

It was a cold December afternoon, about a month after the normal harvest season. A passing motorist wondered why a tractor and corn picker were stopped—but with the machinery still running—in the center of a corn field so late in the year. When he stopped to investigate, he found a farmer caught in the husking rolls of the corn picker. The farmer had been trying to harvest corn, but the husking bed of the picker kept plugging. An hour and a half earlier, he had tried to unclog the machine while it was running and his hand had been caught.

The motorist shut off the tractor and called the local emergency squad. The members were well experienced and knew how to extricate auto crash victims quickly. They assumed it would not take long to remove the farmer from the corn picker. Two of the squad members placed blankets around the farmer, applied chemical heat packs, and lay next to him to try to keep him warm. Other squad members disconnected the PTO driveline and tried to run the machine backwards. The drive mechanism, however, was designed to operate in only one direction; and the husking rolls did not move. (If the machine had been run backward, the farmer might have been further injured.) The squad then checked to see if the rolls could be dismantled or unbolted from the machine. Bearings mounted in two large castings held the rolls, but no one could see how to remove them. So the squad attempted unsuccessfully to pry the rolls apart with a crowbar.

Next they inserted a hydraulic spreading tool between the rubber-covered rolls, but it kept slipping out. Finally, someone cut several rubber sections from the rolls to provide an opening for the tool. This time they could place the rescue tool in a better position. It pried the rolls apart enough to break one roll and relieve the pressure. However, the farmer's hand was still caught. So the crew tried to wedge the rolls apart with a wooden block and move the rescue tool to a new position. But the block began to slip as soon as pressure was taken off the spreading tool; the tool could not be removed without further injuring the farmer's hand.

Other crew members then used a cutting torch to cut away the casting holding the rollers on the side opposite the victim. One member stood by with a CO_2 extinguisher to extinguish any chaff and straw that might catch fire. As soon as the cut casting gave way, the hand was freed. The entire rescue took about an hour and a half.

Chapter 3

Farm Structure Emergencies

Grain Entrapments — On-Farm Storage

This section considers incidents in small on-farm storage structures of less than 25,000 bushels in capacity. See the section on commercial storage (page 38) for information on rescues from larger bins, tanks, or concrete silos.

Introduction

Generally there is little danger from stored grain, but flowing grain presents potential suffocation and entrapment hazards. Grain bins (figure 3.1), which appear relatively safe, have actually been the scene of hundreds of deaths and entrapments. In addition, grain-transport equipment, commercial elevators, and grain-processing facilities have been involved in numerous flowing-grain entrapments.

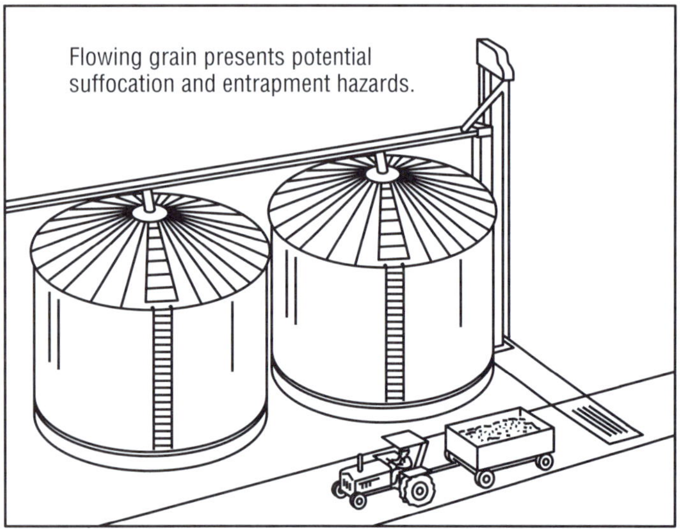

Figure 3.1
Grain bins

Injuries

Most full engulfments of an individual in flowing grain, regardless of the setting, result in a fatality. In one study, over 90 percent of rescue attempts in which the victim had been engulfed beneath the surface of the grain were found to be unsuccessful.

In cases of partial entrapment, however, few injuries have been reported other than dust inhalation. In a very small number of cases, there were injuries to a lower extremity caused by contact with the unloading auger in the base of the bin or silo.

Types of Entrapments

Engulfment in a Flowing Column of Grain

The most common mishap involving grain storage structures is entrapment or suffocation resulting from an individual being drawn into a flowing-grain column. When a grain bin is emptied, the grain flows in a funnel-shaped path above the outlet, which is normally located at the bottom center of the bin. For large grain bins that may have several outlets in the floor, a funnel will form over each outlet. The flow velocity increases as grain flows from the bin wall at the top of the grain mass into a small vertical column at the center of the bin. This vertical grain column flows down through the grain mass at nearly the same rate as the unloading capacity of the unloading auger. Essentially, no grain flows in from the surrounding grain mass.

The rate of grain flow in the center of a grain bin or storage facility is so great that escape is almost impossible once a victim is caught in the flow (figure 3.2). Once engulfed in the flow of grain, a victim is drawn to the discharge outlet of the bin or storage facility within seconds. Entrapment is similar to being drawn into a water whirlpool. In some cases, the victim is drawn into the unloading auger and slows or stops the grain flow.

The few survivors of this type of engulfment have reported that they deliberately covered their mouths and noses with their hands and did not panic. They have also expressed amazement at the tremendous speed of their entrapments.

The condition of the grain also appears to play a major role in the survival of completely submerged victims. It is believed that someone buried in caked or spoiled grain has a greater chance of survival because caked grain masses form open air pockets, which both make breathing possible and reduce the pressure of the grain on the body. That the grain is spoiled, however, is a primary reason why entrapments occur, since workers typically enter the bin to break up caked grain. Spoiled grain also presents hazards to emergency personnel from unstable surfaces and exposure to dust and toxic molds.

Children account for a disproportionate share of entrapments and suffocations in flowing grain. In most cases, a considerable amount of time elapses before they are found in the grain, after which survival is highly unlikely.

Entrapment in Grain-Transport Vehicles

Entrapments and suffocations can also occur in grain-transport vehicles, most often gravity-flow wagons or trucks equipped with grain beds. The victim either becomes buried during loading from a combine or a storage bin or is drawn into the flow of grain as the vehicle is unloaded. Nearly all transport-vehicle entrapment victims are males under sixteen years of age.

Collapse of Horizontally Crusted Grain Surface

Entrapment or suffocation can occur when a victim enters a bin in which the surface of the grain has become crusted due to spoilage. This crust forms a cavity beneath the surface as grain is removed. The victim breaks through the crust and is quickly covered by the avalanche of grain collapsing into the cavity (figure 3.3). Survival in this type of entrapment is less likely, in many instances, because running unloading equipment draws the victim even deeper into the grain mass.

Collapse of Vertically Crusted Grain Surface

Occasionally, victims are buried beneath a collapsed wall of freestanding grain (figure 3.4, page 34). Grain in good condition will pile at a 25–30° angle, but spoiled or caked

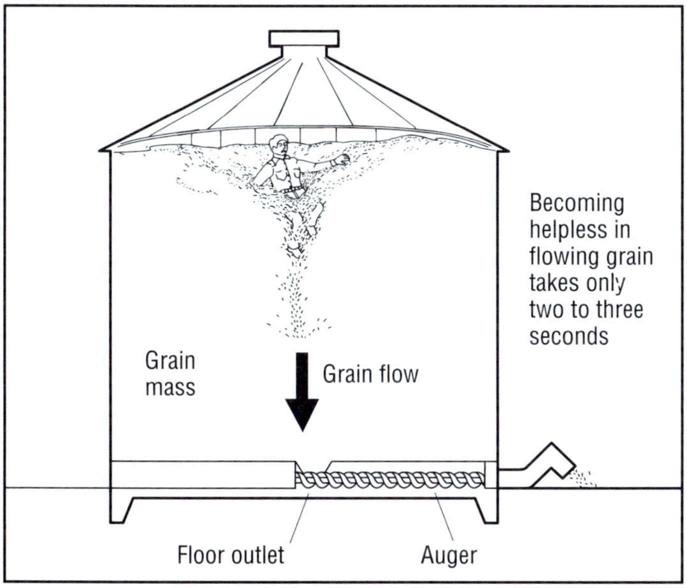

Figure 3.2
Entrapment in flowing grain

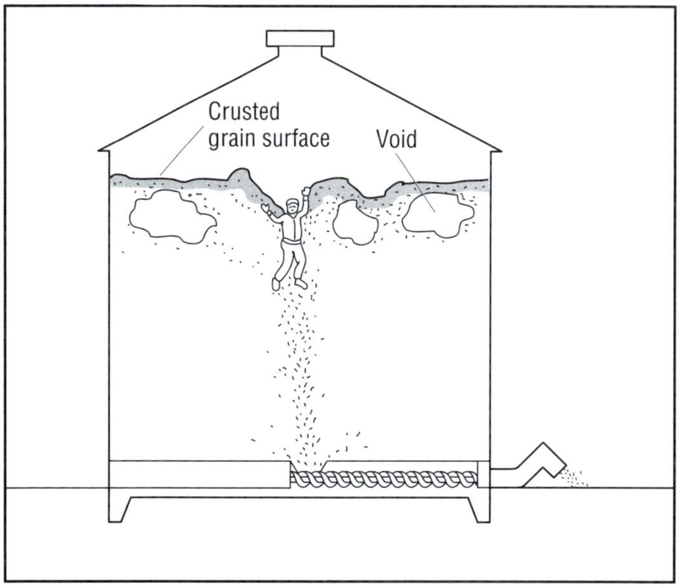

Figure 3.3
Collapse of horizontally crusted grain surface

grain can stand almost vertically. The danger of avalanche and engulfment is very real as a worker removes grain from the base of a vertical pile. Reports of this sort of entrapment generally come not from farm operations but from large grain-handling facilities where large-capacity storage facilities are used or grain is stored outdoors in piles.

Essential Safety Precautions

1. Treat a grain bin in an emergency situation as a confined space, and follow standard procedures for entering a confined space (see sidebar, page 35). Every rescuer entering the grain bin must wear both SCBA (self-contained breathing apparatus) or supplied-air respirator (SAR) and a body harness attached to a lifeline.

2. Generally, the roof of a grain bin may not provide adequate anchors to secure lifelines. In some cases, you may have to work from the ground to ensure secure anchors.

3. Always assume that a victim entrapped or completely submerged in grain is alive. Successful rescues have taken place even when the victim was completely engulfed in 10 feet of grain for over two hours. In other situations, access to the victim could have been expedited if rescuers had assumed that the victim was still alive.

4. Do **not** start the unloading auger (figure 3.5). The victim may be further injured by being drawn deeper into the grain, pulled into the unloading auger, or wedged in the opening.

5. In cases where the victim is partially submerged, stay out of the bin until a rescue plan has been developed. Rescuers entering the bin can disturb the grain surface, causing grain to avalanche on the victim. If the victim needs immediate assistance, it may be necessary to suspend a rescuer above him or her.

6. Never attempt a rescue unless at least three rescuers are participating: two outside—on top of the bin—and one inside (figure 3.6). The rescuer inside the bin should wear a body (parachute) harness and SCBA and be tied with a safety rope to the two rescuers outside. The outside rescuers should be able to lift and pull out the rescuer inside the bin without having to enter it themselves. In some cases, more than two rescuers may be needed outside the bin to provide backup for the rescuers and assist in lifting out the victim.

7. Dust or mold spores from spoiled grain can cause severe allergic reactions in some individuals. Always use SCBA or SAR. For rescuers working outside the bin, paper-like dust masks, called toxic dust-rated masks or NIOSH/MESA-certified dust respirators, are generally sufficient; but such masks should be changed frequently (figure 3.7). **Caution:** If any rescuers begin to exhibit shortness of breath, tightness in the chest, or dizziness, order them to leave the area and keep them under observation. Severe reactions to grain dust or

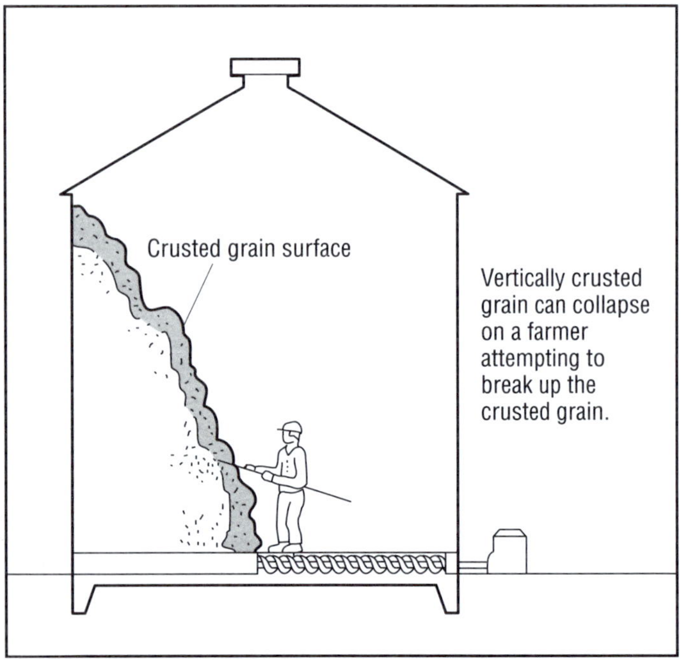

Figure 3.4
Collapse of vertically crusted grain surface

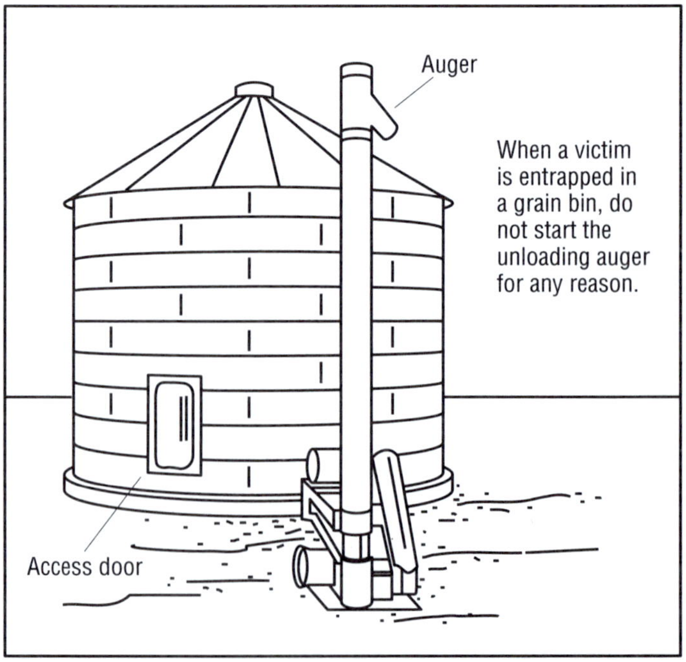

Figure 3.5
Grain bin and unloading auger

Essential Safety Precautions When Entering Confined Spaces

A confined space is a space that has limited means of entry or exit, is large enough for a person to enter, and is not designed for continuous human occupancy. Silos, manure pits, grain bins, and storage tanks are typical confined spaces on farms. Observe the following precautions to avoid potential fatalities of rescuers and victims in emergencies involving confined spaces:

1. Confined spaces should be posted with "KEEP OUT" signs and other placards that alert employees, emergency personnel, and others to potential dangers. Upon arrival at the scene, look for any posted warnings before entering a structure that may be a confined space.

2. Do **not** allow anyone to enter a confined space unless properly trained and equipped with a self-contained breathing apparatus (SCBA) or SAR and a safety harness with a lifeline. Cartridge-type masks are **not** safe, and a person without an SCBA may lose consciousness within seconds from toxic gas exposure or oxygen deficiency.

3. Rescuers should **not** place their own masks on victims or remove their own lifelines. Ropes, harnesses, carriers, and oxygen for victims can be lowered into the space if necessary.

4. The rescuer's lifeline should be attended by at least two people outside the confined space or be connected to a mechanical lifting device.

5. If possible, ventilate a confined space during a rescue with built-in fans or aerators. (Do **not** lower a fan into a confined space; a spark might cause an explosion of accumulated flammable gas.)

6. Prohibit smoking, open flames, or other spark-producing activities throughout the rescue.

7. Victims should be brought out as quickly as possible, given 100-percent medical oxygen, and transported to the hospital for evaluation, even if they have no physical complaints.

8. Rescuers should **not** enter any building where manure is being agitated or emptied. The pump should be turned off in the event of any emergency. Vigorous mechanical ventilation must be provided during agitation and pumping, and for a few hours after pumping has stopped, to allow gases to escape.

(Also refer to table 3.1, "Gases present in agricultural confined spaces," on page 54.)

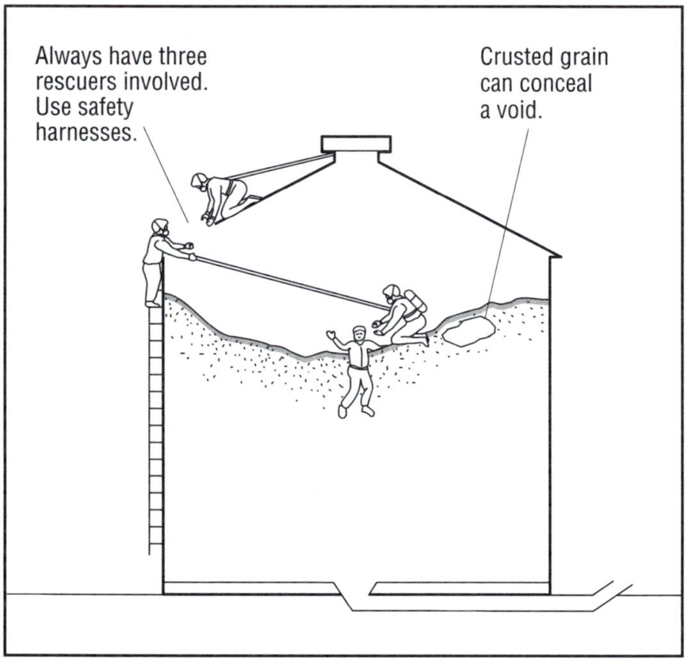

Figure 3.6
Rescuing a partially submerged victim in crusted grain

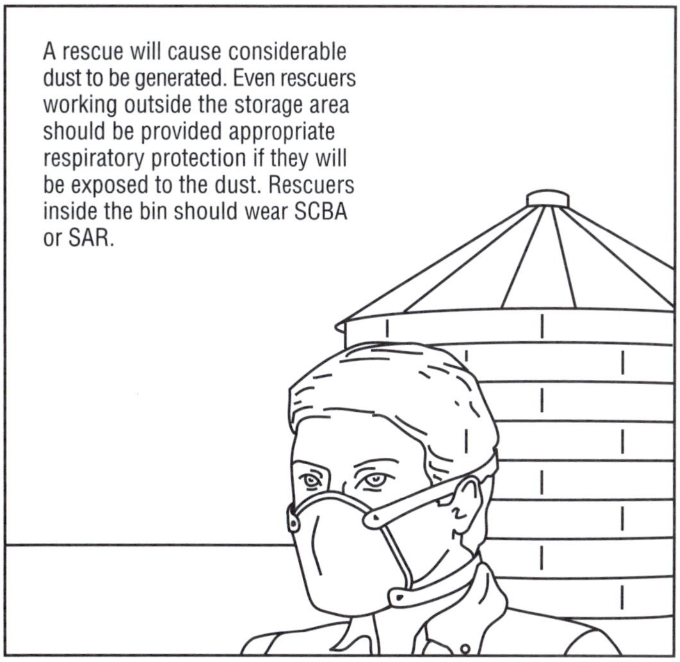

Figure 3.7
Respiratory protection for rescuers working outside the storage area

mold often require hospitalization. Note that such allergic reactions may not surface for several hours. Following the rescue, everyone at the rescue site should shower and change clothing; otherwise, dust and mold will be transported home and can affect rescuers' family members as well.

8. On a hot day, the inside temperature of a grain bin can reach intolerable levels. Rescuers in rescue gear and SCBA need to be monitored for signs of heat stress and rotated if the rescue is prolonged.

Rescue Procedures — Small On-Farm Grain Storage Structures

Rescue Procedures — Completely Submerged Victim

1. If someone is certain how to operate the aeration fan in a grain bin, start the aeration fan to circulate air through the grain as soon as possible in a storage bin rescue. **Caution:** Do **not** start the grain drier. If the aeration fan operates in conjunction with the grain drier, turn off the drier function control at the electrical box. Victims of storage bin incidents generally die of asphyxiation due to blockage of their airways. Reports from several incidents suggest that operating the aeration fan can increase the air supply of a completely submerged victim.

2. Time is critical if the victim is to survive. If the victim is completely submerged, the only solution is to remove grain from the bin as rapidly and carefully as possible (see the following section).

3. Do **not** attempt to use these procedures for large-capacity or commercial storage facilities; there are important differences in rescue procedures (see "Grain Entrapments—Commercial or Large-Capacity Storage Structures," pages 38–39).

4. Efforts to dig out a buried victim from inside the bin have generally failed because of the large amount of grain involved and the tendency of grain to back-flow. In most cases, the victim is already on the floor of the bin, directly over the outlet.

5. It is usually safer and faster to remove the victim through the access door on the bottom of the structure or through an opening cut into the side of the bin after the grain has been removed. Attempting to carry out a rope rescue through the roof opening requires considerable expertise and can be unnecessarily risky and time-consuming.

6. The only successful emergency procedure for rapid grain removal has been to cut large, uniform openings around the base of the bin (figure 3.8). These openings, cut in the metal skin of the bin with abrasive saws, air chisels, or cutting torches, greatly reduce the time needed to access the victim. Do **not** allow anyone to stand on top of the grain as it is being removed.

   Caution: Be sure to have fire protection at hand during all cutting operations, and be alert for fire. Have a charged water line available throughout the rescue. In addition, because of the potentially toxic coatings on metal panels used in bin fabrication, avoid breathing the fumes produced during the cutting operation. Use a cutting torch only when no other cutting equipment is available.

7. If the position of the victim is known, cut the openings directly below the level of his or her feet. If the position of the victim is not known, cut emergency openings 4–6 feet off the ground to reduce the impact of grain piling up outside the bin, which could block the flow of grain. Make cuts within individual panels rather than across panels. V-shaped cuts, 30–40 inches across, form valves which, when bent upward, allow grain to flow freely. When bent back into place, they slow or stop the grain flow.

8. Space the openings uniformly around the bin to re-

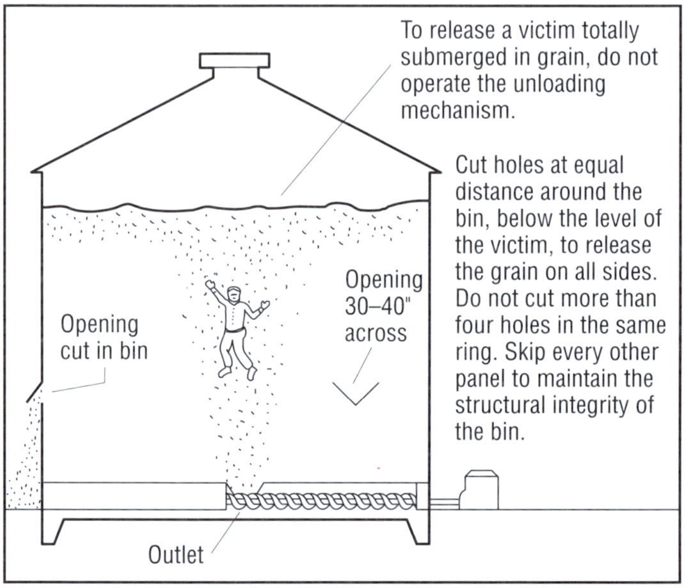

Figure 3.8
Rescuing a completely submerged victim

duce the risk of collapsing or toppling the bin and to make it easier to remove grain from around the base. Cut openings in every other panel or until the victim is uncovered. Do not cut more than four holes in the same ring. Using front-end loaders and/or shovels, move grain away from openings during emptying.

Caution: Do **not** allow anyone to stand on top of the grain as it is being removed.

9. Make observations from the bin hatch or support a rescue worker above the grain with a safety harness. Once the victim is uncovered, the openings in the bin can be closed to allow safe access by rescuers.

10. Only one rescuer is needed inside the bin to assist the patient. He or she should immediately clear the patient's airways and check vital signs. Then he or she should proceed with basic life support or provide oxygen, if needed, and keep the victim calm. A protective barrier may be needed to keep additional grain from in-flowing on the victim. (See numbers 3 and 4 of "Rescue Procedures—Partially Submerged Victim" below.)

Rescue Procedures — Partially Submerged Victim

1. If possible, lower a rescue squad member into the bin to reassure the victim and to try to attach a harness or lifeline (figure 3.9). Do **not** try to pull the victim free using ropes or harnesses. The tremendous drag created by the grain is likely to cause further injury. Use the lifeline only to reassure the victim and prevent further sinking.

2. Check the victim's airway for lodged grain and administer oxygen if he or she experiences difficulty breathing. Sometimes panic and the exertion of trying to work free of the grain, rather than pressure on the chest and lungs, cause the victim to have difficulty breathing.

3. Fabricate a coffer dam around the victim using whatever resources are available. A coffer dam could be made of strips of plywood, metal roofing, or even old tires positioned one on top of the other. It may be necessary to enlarge the roof opening to get the coffer dam material into the bin.

4. One approach that has been developed and used successfully at some commercial facilities is a grain rescue tube (figure 3.10). This tube is an aluminum cylinder that is 5 feet long and 3 feet across. It consists of three, 120°-angle sections that can be moved through grain bin hatchways as narrow as 33 inches. Exterior handles enable rescuers to properly position the tube around the victim and force the tube down into the grain. Interior handles provide hand holds for the trapped victim. In addition, a removable top protects the victim from further grain flow and provides valuable air space. This tube may be difficult to use if the

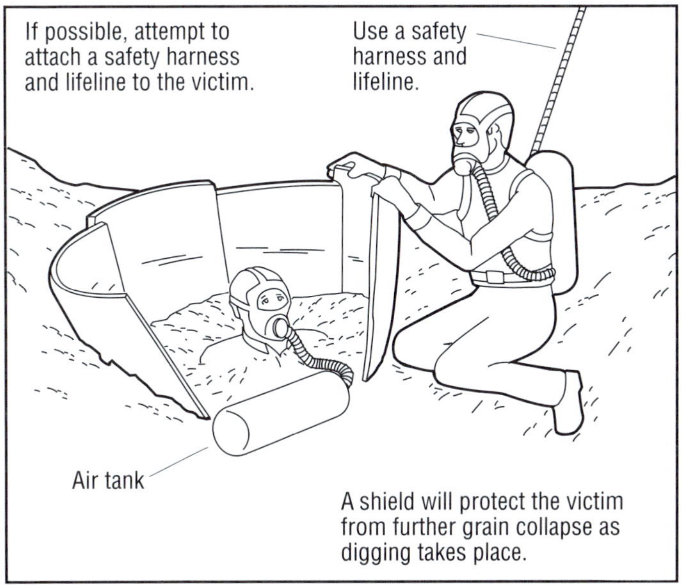

Figure 3.9
Rescuing a partially submerged victim

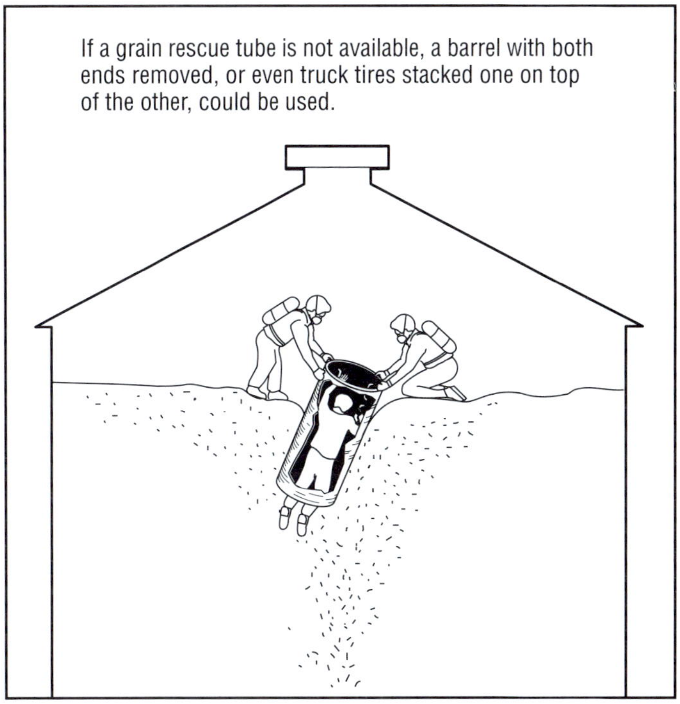

Figure 3.10
Grain rescue tube

victim's body is not perpendicular to the grain surface or if a limb is extended into the grain.

5. If the victim is buried only up to the waist, it might be possible to dig him or her out using the coffer dam or grain rescue tube and a secured safety line and harness. Use a board, sheet of plywood, or short ladder as a work platform. The board or ladder will distribute weight over a larger area.

6. A victim buried above the waist is in serious danger, and grain should be removed from the bin quickly by cutting openings as described in the previous section. Efforts to dig out deeply buried victims have proven very difficult.

Rescue from Grain Transports

Entrapments in grain-transport vehicles usually present less complicated rescues than those in large storage facilities. Transport entrapments, however, are often fatal because almost all involve children. As with storage facilities, do **not** use the gravity-flow opening on gravity-unloading vehicles unless the victim is completely submerged. Partially submerged victims, especially children, are apt to be drawn deeper into the grain or become wedged in the opening. If the victim is completely submerged, open the gate and reach up into the flowing grain to attempt to feed or pull him or her through the opening or at least expose the head and airways.

One possible procedure with gravity wagons is to flip the wagon on its side, spilling out both the grain and the victim. The wagon should be tipped in the opposite direction than that from which the grain is being unloaded. Speed is critical since the potential for a child to survive under even a thin layer of grain is not very good. Be aware that tipping the wagon may further injure the victim.

Grain Entrapments — Commercial or Large-Capacity Storage Structures

Over the past few years, the average size of both on-farm and commercial grain storage structures has increased significantly. Units of 100,000 and even 500,000-bushel capacity are now common throughout major grain-growing regions (figure 3.11). The strategies for rescue from smaller on-farm storage bins are generally not appropriate for these larger structures. In fact, attempting to cut open one of the larger steel bins or welded tanks will most likely result in catastrophic structural failure, because of the tremendous pressures against the walls of the structure.

The nature of entrapments in large grain-storage structures is similar to that of entrapments described in the previous section. The primary differences are the size of the structures and the volumes of grain involved. The possibility of escape or rescue from a commercial facility is highly unlikely. In fact, over 90 percent of the victims who are fully engulfed do not survive.

Essential Safety Precautions

Caution: Entrapments in large grain-storage facilities have on several occasions resulted in multiple deaths because rescuers attempted to carry out the rescue with inappropriate equipment or inadequate training or assistance. Never rush to attempt a rescue in an entrapment situation before a well-thought-out plan has been developed. Then, be sure to follow these essential safety precautions:

1. STOP! Never rush into an entrapment situation in an attempt to rescue the victim. There could be hazardous conditions present, such as cliffed grain or toxic environments, that could lead to the rescuer becoming another victim. Treat the facility as a confined space and use appropriate life support and rescue equipment (see sidebar, page 35).

2. Shut down and lock out all unloading equipment (figure 2.32, page 22).

3. Activate the appropriate emergency services that will be needed to conduct a successful and safe rescue. This includes the local high-angle rescue team, first-re-

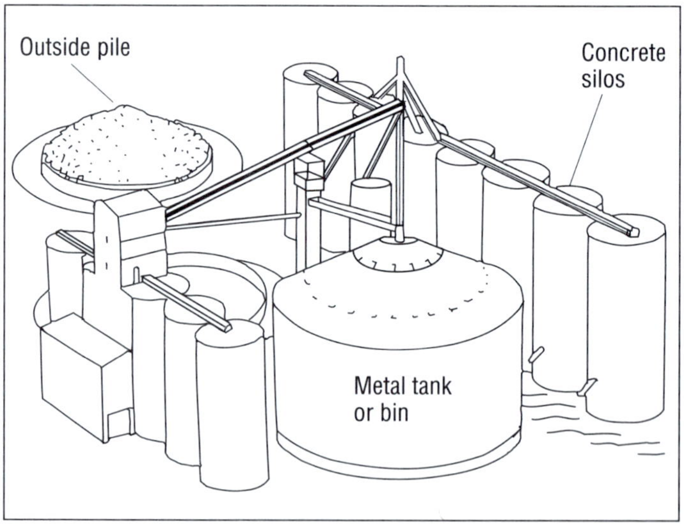

Figure 3.11
Large-capacity grain storage facility

sponse personnel from the plant, emergency medical services, and law enforcement to provide a secure area.

4. Turn on aeration and roof exhaust fans to remove suspended dust at the top of the structure and to help provide a source of fresh air. This may prove very important in cases of full engulfment.

5. Assemble all employees at a predetermined location to confirm the whereabouts of every employee and determine who and how many may be entrapped.

6. Assess the situation carefully to identify such hazards to rescuers as instability of the grain mass or dangerous atmospheric conditions inside the structure and its access points. Be aware of structural hazards, including the potential for falls, contact with overhead power lines, and the presence of pesticide residues and spoiled grain. If there is any question, test for toxic conditions.

7. Implement a situation-specific action plan. Use only trained and experienced personnel for rescues involving confined-space or high-rise situations.

Rescue Procedures — Large-Capacity Grain Storage Structures

Partially Submerged Victim

Procedures for rescuing a **partially** submerged patient in a large grain storage structure are somewhat comparable to those used in a small, on-farm bin, with one extremely important exception. Do **not** jeopardize the integrity of the structure by attempting to cut openings in the structure; this action could result in a catastrophic structural failure that would endanger rescuers.

The victim cannot be forcefully pulled out of the grain mass without suffering additional injuries. He or she must be dug out. Use a rescue tube if available or fabricate a coffer dam and dig out the victim. (See numbers 3 and 4 of "Rescue Procedures—Partially Submerged Victim" for small on-farm storage structures, pages 37–38).

Completely Submerged Victim

In a large structure loaded with grain, the only strategy that has been documented as successful in rescuing a **completely** submerged victim alive is to remove him or her through the bottom draw-off or outlet when this is physically possible. The possibility of successfully rescuing a fully submerged victim is, however, very small.

Real-Life Scenario

Large chunks of spoiled grain blocked the grain flow from a 10,000-bushel, steel storage bin. A farmer left the unloading auger running and entered the bin to break up the grain with a long pipe. As he moved across the grain surface, the blockage suddenly broke loose. The grain abruptly resumed flowing, and the farmer was drawn into the flow of grain. He had time to scream for help only two or three times before being completely buried. An assistant who heard the shouts shut off the auger. He raced to the top of the bin but could not see the farmer. He immediately called emergency rescue units.

Once on the scene, emergency personnel used a torch to quickly cut three, 3-foot-wide, V-shaped openings around the circumference of the bin. They left the top portion of the V connected to the storage bin to form a control valve.

The rescuers requested additional help from nearby units to move grain away from the base of the bin. They also used a tractor-mounted front-end loader to keep the grain flowing freely. As the level of the grain in the bin lowered, two rescuers using lifelines and harnesses were lowered to the center of the bin with shovels to probe for the victim. Each rescuer was secured by two or three backup rescuers, who helped him avoid being drawn into the rapidly flowing grain as it was emptied from the bin.

As soon as they uncovered the victim's head, they cleared the grain from his mouth and nostrils and administered oxygen. Rescuers constructed a plywood shield around the victim's head to prevent additional grain from flowing in on him. They attached an additional lifeline to the victim as soon as they could. Once he was freed from the remaining grain, rescuers removed him on a Stokes litter through one of the openings they had cut in the side of the bin.

The victim survived the incident. It was believed that the chunks of crusted grain had provided ample air pockets around him, allowing him to breathe.

Silo Incidents — Gas Exposures, Unloader Entrapments, and Fires

The cylindrical vertical silo, or tower silo, preserves and stores livestock feed. Two types of vertical silos are common: conventional silos and oxygen-limiting silos. Silos vary in circumference, height, type of construction, and design. Construction can be either poured concrete or concrete stave, steel, or (in older silos) wooden stave or tile. Some have roofs or caps, while some conventional silos may be open at the top. Silos have all the characteristics of a confined space; and rescuers must follow confined-space procedures when entering silos (see sidebar, page 35).

A conventional silo (figure 3.12) may have an electrically operated unloading device that hangs by a cable down to the surface of the silage. The unloader rotates inside the silo on the surface of the silage, chops off a thin layer of silage, and blows it down a closed chute to deliver to livestock. This chute covers a series of doors on the side of the silo (figure 3.13). Conventional silos can also be equipped with a bottom unloading system and not have an unloading chute or with a top unloading system using a tube formed in the center of the silo during filling.

Oxygen-limiting silos (figure 3.14) will be encountered on some farms. Oxygen-limiting silos have an unloading mechanism located near the bottom and will not have an unloading chute. They also have a small vent or airtight hatch cover at the top which can be closed to exclude air. The oxygen-limiting silo is a structure designed specifically to minimize the amount of oxygen in the silage. To

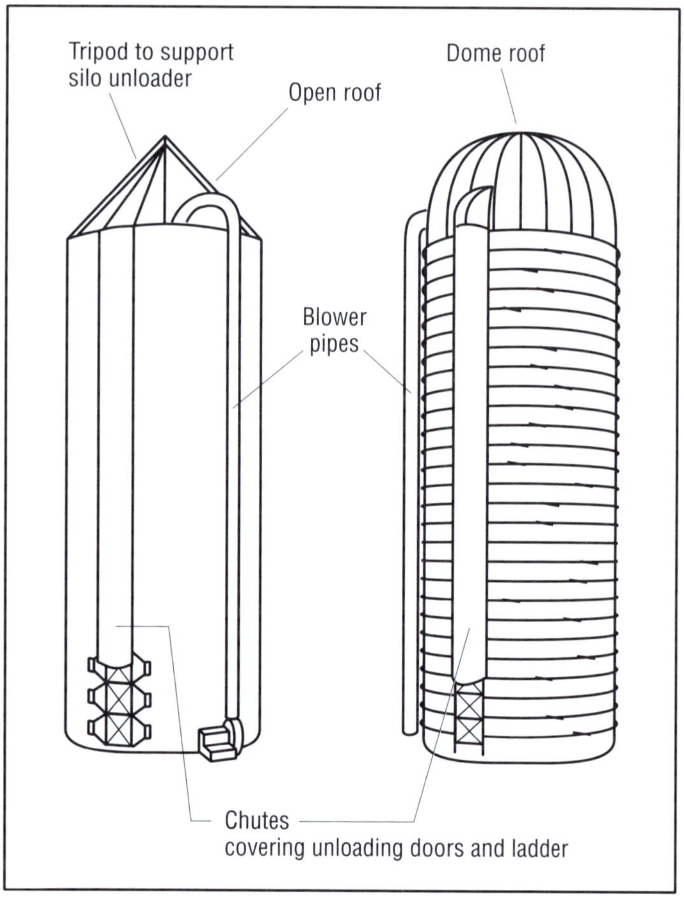

Figure 3.12
Conventional silos

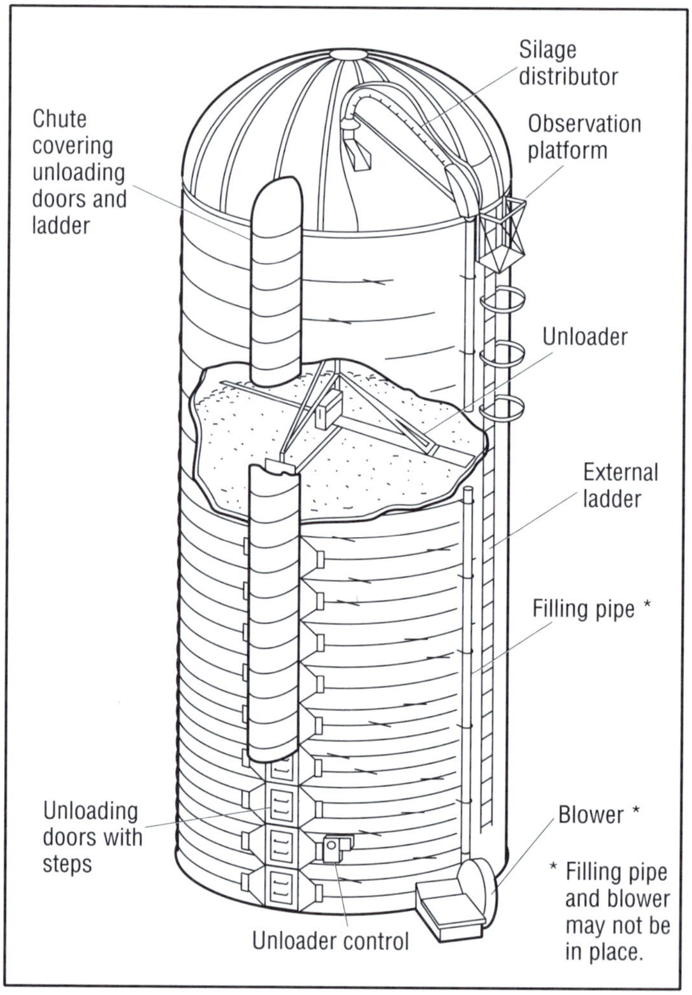

Figure 3.13
Top-loading conventional silo

enter safely, it is necessary to follow confined-space procedures, including wearing a self-contained breathing apparatus (SCBA). Oxygen-limiting silos may be entered through the roof access door or the bottom access door used to service the unloading mechanism.

Roof construction is a good indicator of silo type. Roofs on conventional silos are either fully open (have no roof) or have roofs constructed of fiberglass or light metal. Furthermore, conventional silos will have openings in the roof for the silo blower tube and for viewing the silage level. Most also have ladders, as shown in figure 3.13. The roofs of oxygen-limiting silos have an airtight hatch cover and one or more venting valves (figure 3.15).

Some oxygen-limiting silos have been retrofitted such that they no longer minimize the amount of oxygen in the silage. They are referred to as modified oxygen-limiting silos. Even though they have been modified, they should be treated as oxygen-limiting silos during rescues or fires.

Conventional Silo Incidents

For conventional silos, three types of incidents may require a victim's rescue: exposure to silo gas, entrapment in the silo unloader, or a medical emergency in the silo such as a heart attack. These incidents occur when the victim enters the silo, to work on equipment or break up silage, or accidentally falls into the silo. For each situation, the victim may not be able to descend the enclosed chute ladder and will require extrication using a litter, backboard, body sling, or harness.

If necessary, the silo unloader in a conventional silo may be raised by the cable and hoist mechanism, although there is seldom a need to do this. The unloader discharge spout will block one door, but there are other doors close to the silage surface that can be opened.

Nearly all silos have an external ladder (on the side opposite the chute) used to attach the filling pipe from a silage blower. These ladders normally start about 8–10 feet from the ground to prevent access by children.

Caution: External ladders on silos vary considerably in style of construction, condition, and inclusion of safety features. If there is any question concerning the safety of the ladder, other means should be used to reach the top of the silo. On a conventional top-unloading silo, the unloading doors, when in place, provide a ladder inside the unloading chute. These door "handles" can be slippery or damaged; they should be used with care.

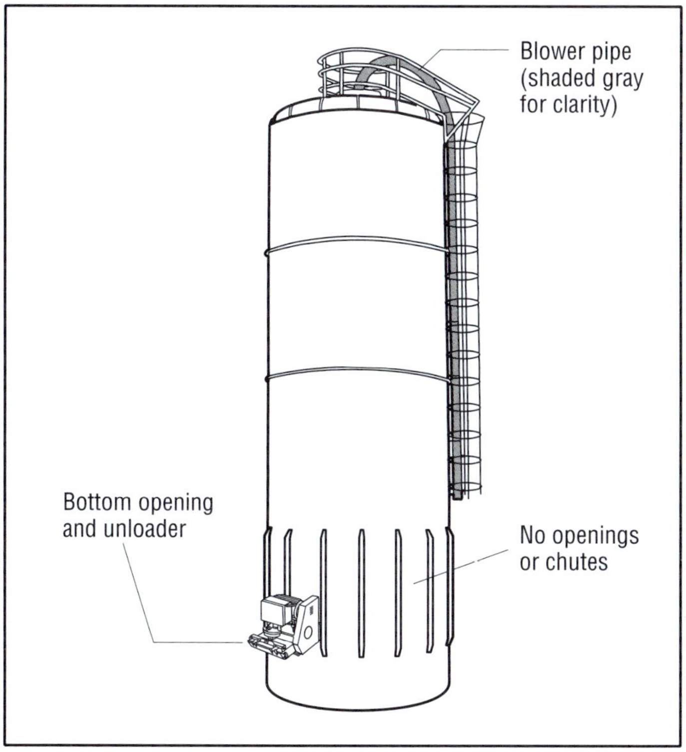

Figure 3.14
Oxygen-limiting silo
Adapted with permission from A. O. Smith Engineered Storage Products Company. Note: The term "oxygen-limiting silo" is not preferred by A. O. Smith. They prefer "bolted-together steel silo" or "closed steel silo."

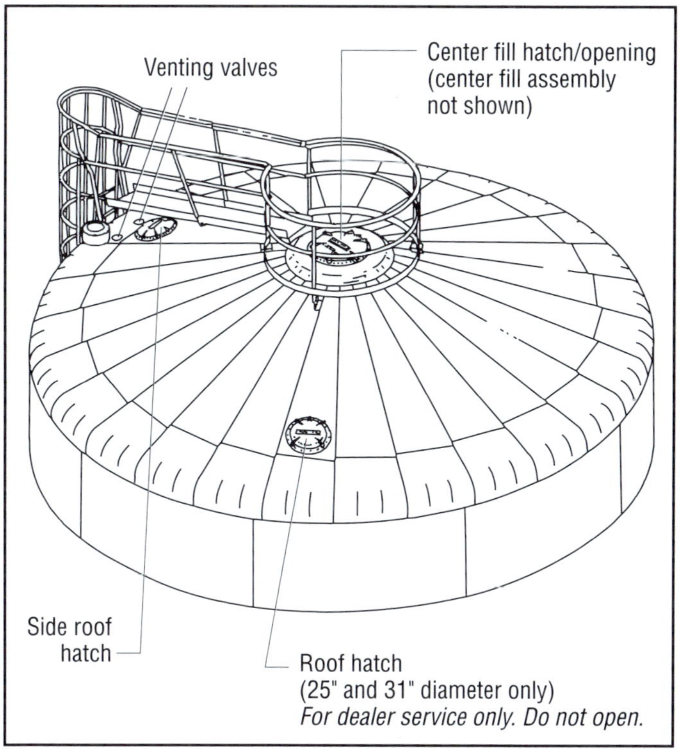

Figure 3.15
Roof of an oxygen-limiting silo
Adapted with permission from A. O. Smith Engineered Storage Products Company. Note: The term "oxygen-limiting silo" is not preferred by A. O. Smith. They prefer "bolted-together steel silo" or "closed steel silo."

Silo Gas Exposures

Silage is formed by natural chemical fermentation. Fermentation begins to occur in the chopped forage within a few hours after it is placed in the silo. A variety of gases, including carbon dioxide and nitrogen dioxide, are released as the silage ferments. Carbon dioxide, the primary component of silo gas, can reach concentrations of 30 percent or higher and cause suffocation.

Although it is only one of a number of gases that are released by silage, many people refer to nitrogen dioxide as "silo gas." Even in very low concentrations, nitrogen dioxide forms an acid in the lung tissue when combined with moisture, leading to permanent lung damage. Because it is heavier than air, nitrogen dioxide tends to collect at the surface of the silage and flow down the silo chute into adjoining feed rooms or low areas around the base of the silo. In high concentrations, this gas has a strong bleach odor and may appear in the silo-filling area as low-lying yellow, red, or dark brown fumes. In addition to being toxic, it also replaces oxygen and can cause asphyxiation.

Silo gases can remain a hazard for two or three weeks after crops have been put into the silo. The greatest danger, however, especially with corn, occurs one to three days after filling. Although it is unlikely that more silo gas will be produced two or three weeks after filling, a potential hazard always exists—usually when the silo is first opened for unloading.

Silo gas incidents may also occur with oxygen-limiting silos. The health impacts and patient-care recommendations are the same for both types of silos. The rescue procedures described in this section apply to conventional silos only.

Caution: All rescue personnel in or around silos must follow confined-space procedures. Self-contained breathing apparatus (SCBA) is required, regardless of the season or type of rescue. Use extreme caution when rescuing victims in and around silos that have been recently filled.

Injuries

Although the victim of silo gas may feel little discomfort immediately, silo gas causes severe irritation of the upper respiratory tract and may cause potentially fatal lung inflammation (figure 3.16). When silo gas or nitrogen dioxide comes into contact with moisture in the lungs or upper respiratory tract, nitric acid is formed, which can severely irritate the lining of the respiratory tract. A person may inhale silo gas for a short time and notice no ill effects but may go to bed that evening and die due to pulmonary edema

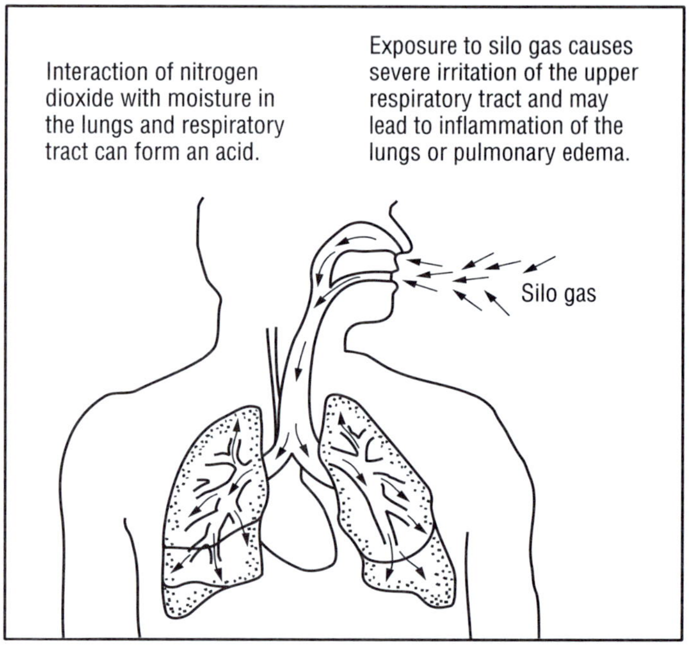

Figure 3.16
Inhalation of silo gas

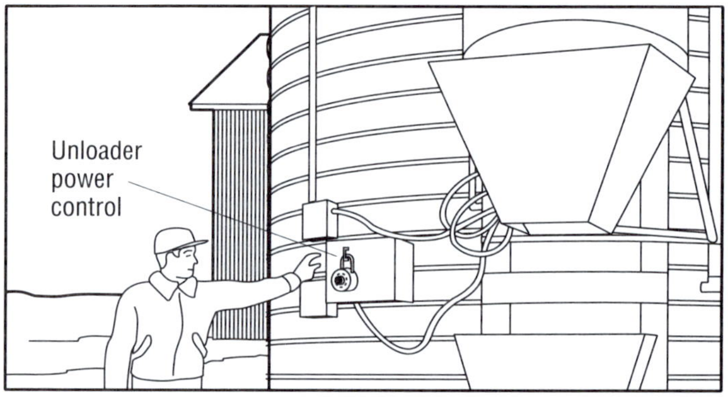

Figure 3.17
Lock out the main power control for the unloader

(fluid buildup in the lungs). A relapse, with symptoms similar to pneumonia, often occurs one to two weeks after initial recovery from the exposure.

Most people who develop initial poisoning symptoms also develop further problems. Therefore, it is extremely important that victims of silo gas exposure seek immediate medical attention—regardless of the apparent severity of the symptoms. Check with your poison control center for proper medical advice.

Rescue Procedures — Essential Safety Precautions

1. Use caution when rescuing victims in and around silos, especially in cases where the silos were filled within the past three weeks. Assume that silo gas is present and that there is insufficient oxygen. Use self-

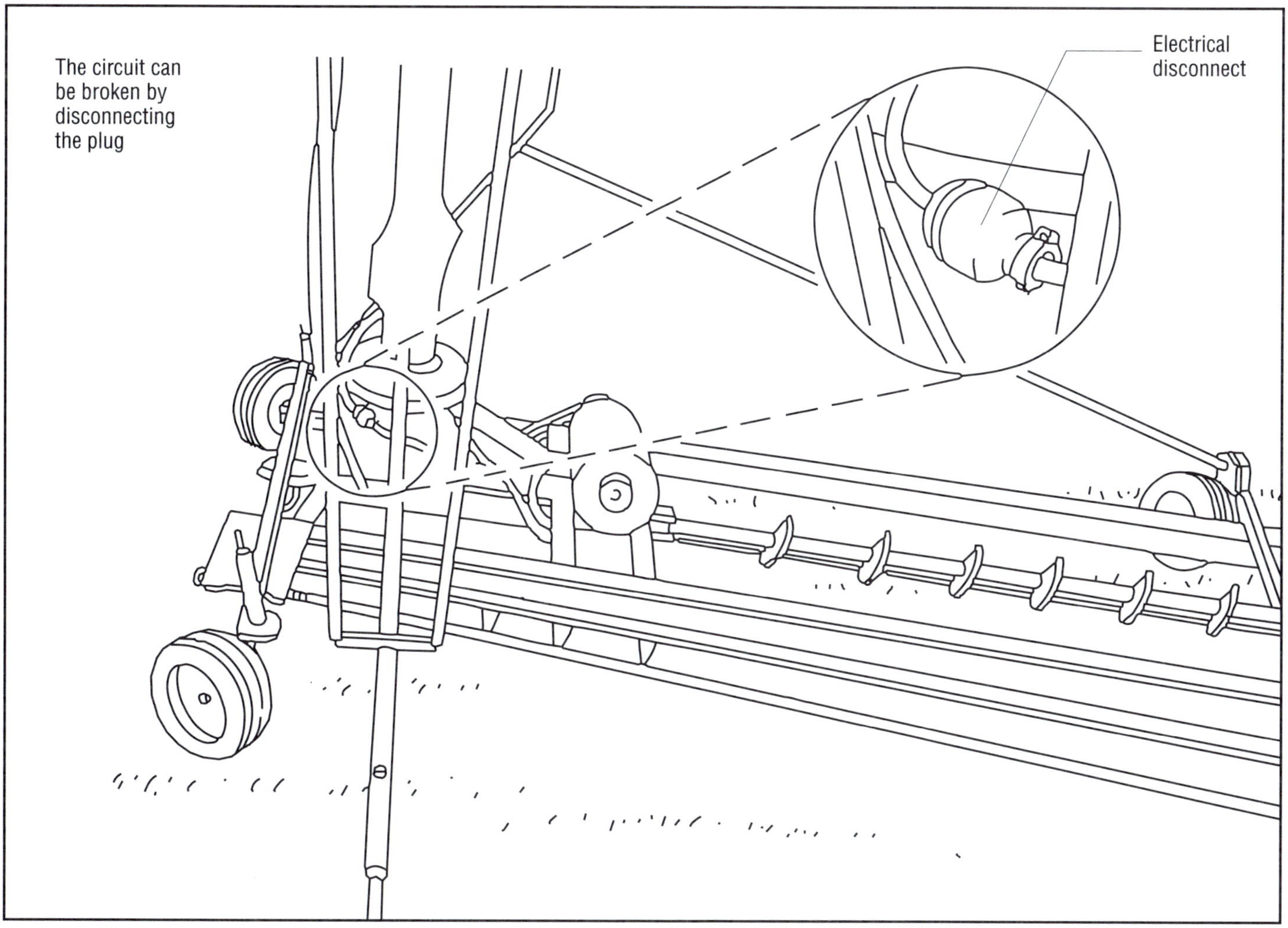

Figure 3.18
Silo unloader

contained breathing apparatus (SCBA) or supplied-air respirator (SAR) when entering the silo, regardless of the season or type of rescue. Use air monitoring equipment to tell which gases are present inside the silo and in what concentrations. This information may be necessary for appropriate medical care of the victim. If a rescuer is unable to climb the enclosed silo chute with SCBA, he or she should use an air line while climbing the chute. SCBA should be attached to tag lines and pulled up the chute once the rescuer has entered the silo with the air line. If an air line is not available, rescuers may have to climb the silo from the outside and be lowered in through the roof opening with SCBA.

Warning: Cartridge respirators do **not** provide sufficient protection from silo gas. SCBA or SAR must be worn.

2. Locate and lock out the main power control for the unloader. Make sure the power is off to ensure that the unloader is not started during the rescue. Lock out the power supply (as shown in figure 3.17, previous page). If the switch cannot be locked out, have someone guard it or tape over it. Once inside the silo, unplug the unloader motor (figure 3.18).

3. If the silo is equipped with a silage blower, turn it on before entering and keep it running as long as the victim or rescuers are in the silo. This will help to dispel any gases present at the surface of the silage. However, the blower will not be effective if the silo is less than three-quarters full.

4. All emergency personnel entering a silo in which they suspect silo gas is present must use lifelines with body harnesses and work according to other emergency confined-space entry procedures (see sidebar, page 35).

5. Be cautious on silo ladders and use a life belt and hook. The rungs of the enclosed chute ladder double as door

handles and do not provide secure footing. Some may be broken, loose, or missing. External ladders may also be missing or weakened because of weather or mechanical damage.

6. Immediately relieve from duty all rescue workers exposed to silo gas and make sure they receive medical attention. Any rescue worker exposed to silo gas should be seen by a physician. Silo gas-related respiratory problems are sometimes difficult to diagnose. Accumulation of fluid in the lungs and a burning sensation in the chest, nose, or throat are warning symptoms that should be taken very seriously.

7. Silo chute doors should be opened while standing above the door. This will prevent low-lying gas from flowing directly into the rescuer's face. When the doors are removed or opened, silo gas could also flow out the door and down the chute and expose other rescuers. Therefore, all rescuers in the immediate vicinity of the chute opening should wear SCBA.

8. In addition to silo gas, there may be a substantial amount of moldy silage present that will release mold spores if disturbed during the rescue. When inhaled, these spores and dust can cause severe allergic reactions in some people. Respiratory protection should be worn by those with a history of allergic reactions to dust and mold and by anyone else who is handling the silage. For those inside the silo, SCBA will provide necessary protection; while for those outside who may be exposed to dusty conditions, NIOSH/MESA-certified dust respirators are sufficient.

Treating the Victim of Silo Gas Exposure

1. Provide immediate medical attention if the patient is found lying on the silage surface. If possible, lift the patient away from the surface, where gas concentration is greatest.

2. Put an extra self-contained breathing apparatus (SCBA) on the patient if at all possible. Try to move air or 100-percent oxygen into the victim's lungs.

3. Move the patient to fresh air as quickly as possible.

4. Transport the patient to a hospital for medical evaluation.

Real-Life Scenario

A healthy 16-year-old boy was helping fill a conventional silo with corn silage. In the process of topping off the silo, he entered the top of the silo briefly to level off the pile of silage directly beneath the distributor. When he finished, he left the silo and helped complete the evening milking, although he noticed a tingling sensation in his chest. He again noted the burning sensation after dinner, but he went out to a movie, returned home late, and went to bed. His parents found him dead the next morning. He had died in his sleep from pulmonary edema, or fluid buildup in his lungs, that had been caused by silo gas irritation.

Silo Unloader Entrapments

Essential Safety Precautions

When rescuing someone trapped in silage unloading equipment (figure 3.18, page 43), keep these points in mind:

1. All rescue personnel entering the silo must follow confined-space entry procedures including the use of a SCBA or SAR (see sidebar, page 35).

2. The surface of silage is *usually* firm enough to walk on; but, as always, exercise the proper precautions, such as placing a plank or piece of plywood on the surface as a work platform.

3. Silage adhering to the silo walls could fall and further injure the victim or rescuer. In addition, loose doors above the level of the silage could fall on the rescuer.

4. You may have to lift a seriously injured person over the top of the silo using a rope, pulleys, and a backboard, litter, or body sling. If possible, contact the nearest available high-angle rescue team for assistance in removing the victim.

5. Rescue teams should know the height of the tallest silo in the area and be equipped with enough rope to lift a victim out from any location within the silo.

Rescue Procedures — Silo Unloader and Silo Entrapments

1. Use lock-out/tag-out procedures with a padlock (figure 2.32, page 22). Locate the main power control for the unloader. Make sure the power is off, and padlock it before entering the silo (figure 3.17, page 42). If you cannot lock out the switch, have someone guard it or tape over it. Be certain it is kept off during any rescue attempt from either the silo or the unloading mechanism. This protects against both sudden start-up of the unloader mechanism and unintentional contact with any live components. Unplug the unloader motor from its extension cord. (If a padlock is not used or if you are unsure whether the power has been locked out, the power cord should *always* be unplugged at the unloader.)

2. Set up sufficient lighting both inside and outside the silo during night rescues. A portable lighting system is extremely helpful because of the low light levels in most silos. Unusually tall silos may be difficult to light.

3. If the victim is caught in the unloading mechanism, study the equipment to plan the best extrication method. You will have to hoist tools on tag lines up the enclosed chute or drop them by tag line from the top of the silo.

Organizing a Silo Rescue

1. Find out or confirm the height of the silo and be sure that enough rope or line is available to remove a patient from any location within the silo from over the top of the silo, if a high-angle rescue becomes necessary.

2. Prepare to remove the victim from inside a silo with two rescue teams: one operating outside the silo and one inside. Use two-way radio communication. Do not attempt a silo rescue unless sufficient support personnel are available. Remember, not all personnel are trained or comfortable working with high-angle or confined-space rescue.

3. The rescue may be coordinated by one person stationed at the top of the silo. This vantage point gives visual and verbal access (via two-way radio) to teams working inside and outside the silo.

4. The inside rescue team should secure the patient in a Stokes basket, litter, or body sling, depending on the injury or situation.

Primary Options for Removing the Victim

1. If the patient does not need to be secured to a backboard, the best way to rescue is to attach a full body harness and lower him or her down the chute. Have one rescuer equipped with a lifeline and SCBA go down the chute ahead of the patient to prevent the patient from being caught by any obstructions.

2. If access is available to the side of the silo near the silage level, the silo wall may be cut so that the victim can be extricated out through the wall and down. To avoid jeopardizing the structural integrity of the silo, make a hole no larger than is necessary to fit the litter or Stokes basket.

3. If the victim is on a backboard, cut a hole in the outer wall of the chute and pass the victim out the silo door and across the chute to the outside. Before lowering the patient to the ground, tie off the pulley on the silo rings or hardware above the door from which you exited. (Ladder or snorkel trucks can also be used for lowering victims to the ground.)

Other Options for Removing the Victim

In some cases, it will not be possible to take a patient safely down the enclosed chute, especially if he or she has been seriously injured. The outside team should prepare to lift the patient up from the inside and down the outside of the silo in the following manner.

Caution: Not all rescuers or fire service personnel are trained in high-angle rescue or confined-space rescue. If they are unfamiliar with these practices, contact the nearest high-angle or confined-space rescue team. This could be pre-planned into rescue procedures prior to an incident.

1. Securely lash a ladder to the outside of the silo structure so that it extends at least 6 feet above the silo rim (figure 3.19). **Note:** Before lashing the ladder to the top of the silo, rig a pulley or snatch block to the top of the ladder and run the rescue line through it. The rungs of the silo's external ladder provide a possible place to secure the rescue ladder.

2. Drop part of the rescue line inside the silo to the rescue team there. Then drop the remainder of the rescue

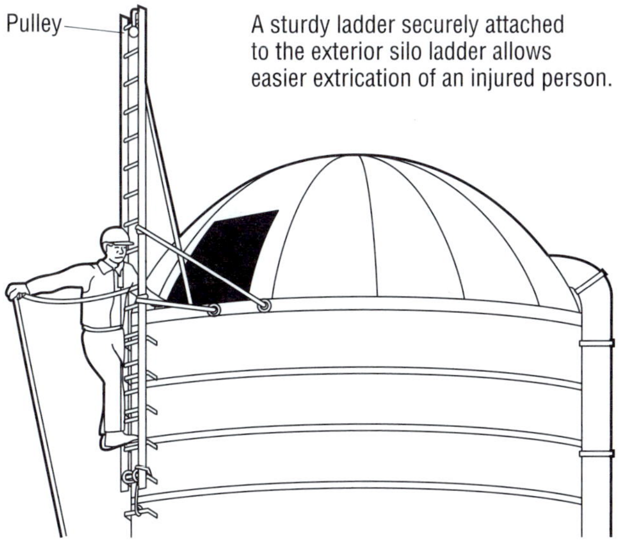

Note: Not all rescuers or fire service personnel are trained in high-angle rescue or confined-space rescue. If they are unfamiliar with these practices, contact the nearest high-angle or confined-space rescue team. This could be pre-planned into rescue procedures prior to an incident.

Figure 3.19
Attaching a rescue ladder to a silo

line to the ground.

3. Attach the line to the patient.

4. Rig another pulley or snatch-block to the base of the silo directly under the ladder lashed onto the top. Run the rescue line through this pulley and to the rescuers on the ground.

5. Most silos have an access door or opening in the top cap or roof; it is either at or near the outside ladder. If this opening is not large enough or if the victim cannot pass through easily, enlarge it or cut a new opening with an air chisel or power saw. Most conventional silo roofs are made of aluminum or fiberglass. You can use wrecking bars or a power saw to tear apart a wood roof. Be careful when working, since weather may have rotted or weakened wood structures. Rescuers at the base of the silo should wear protective head gear to prevent injury from falling debris. The patient needs to be protected from falling objects as well.

6. Assist and secure the victim's ascent from inside the silo and his or her descent to the ground with two tag lines fastened to the litter. Pull the patient up to the top of the silo. A rescuer at the top can then maneuver the litter from inside to outside. Finally, start the descent to the ground (figure 3.20).

Real-Life Scenario

While troubleshooting a malfunctioning silo unloader, a farmer left the unloader running and climbed to the top of the silo to observe its operation. He identified the problem; and then, rather than taking time to climb down and shut off the power, he entered the silo to repair the unloader.

When he stepped over the drive shaft of the unloader's traction wheel, his coveralls caught on a bolt protruding from the shaft. The slowly rotating shaft tightly entangled the farmer's leg and injured him severely. Fortunately, he was dragged within reach of the electric cord supplying power to the unloader's motor, and he was able to unplug the machine.

More than an hour passed before the farmer was discovered by his son, who immediately called for emergency help. Both a fire department rescue squad and emergency medical unit responded to the call. When they saw the location of the victim and assessed the extent of his injuries, they quickly summoned additional help.

After rescuers freed the farmer from the unloader and administered initial emergency medical care, they tried to lower him through the silo access chute. However, the size of the silo opening and exterior covering over the chute made this approach impossible without extensive and time-consuming cutting through the silo structure.

Because they were experienced in high-rise rescues, the rescue team chose the strategy of removing a portion of the silo's cap (roof). Rescuers securely fastened ladders to both the inside and outside top of the silo to act as a suspension point. Two teams of rescuers, one on the ground and the other inside the silo, raised the injured farmer in a basket stretcher to the top of the silo and lowered him down the outside. They used tag lines to keep the stretcher from spinning as it was raised and lowered.

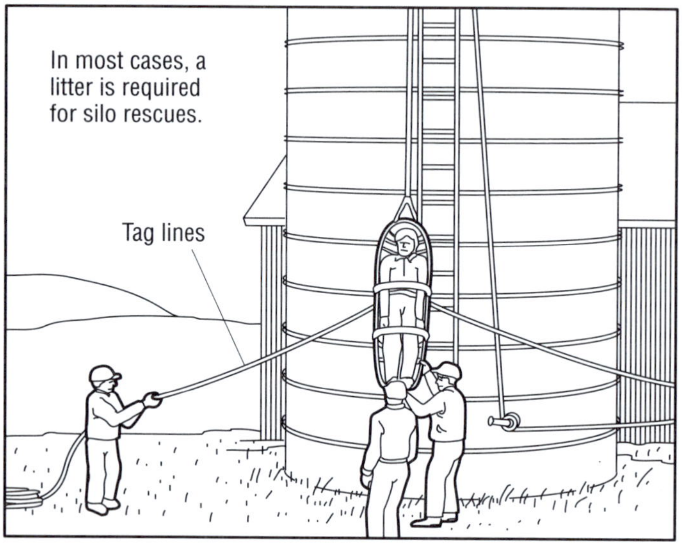

Figure 3.20
Removing a patient from a silo

Silo and Hay Mow Fires

The procedures for fighting fires in conventional silos, oxygen-limiting silos, and hay mows are each different. Using the wrong procedures when fighting fires can be dangerous. Firefighters should be confident about the type of storage before initiating fire-fighting procedures.

Some oxygen-limiting silos have been retrofitted with an unloading system and structural changes that result in the silo losing its oxygen-limiting status. These silos are referred to as modified oxygen-limiting silos. Because of the tight construction of these structures, firefighters must use procedures for oxygen-limiting silos. Water should not be used when fighting fires in a modified oxygen-limiting fire. Rescuers should ask the owner if they are unsure of the silo type.

What Causes Silo Fires?

Plant material continues to respire or "breathe" for a short time after it is cut. After a silo is filled, aerobic respiration produces heat until the oxygen in the material is consumed; then anaerobic fermentation produces heat and preservative acids until the chopped forage becomes stable.

If forage is stored at the recommended moisture content, the water in the forage conducts heat away from the silage mass, and overheating will not occur. If forage is too dry, heat cannot be dissipated quickly enough, and the internal temperature rises until spontaneous combustion occurs (figure 3.21).

As the temperature rises above 130°F, a chemical reaction occurs that may sustain itself. Heat kills microorganisms at 250–400°F and begins to break down the forage by an oxidation process known as pyrolysis. As pyrolysis continues, oxygen within the silage supports a smoldering fire. If the surrounding silage cannot support combustion, the fire may die, leaving a charred cavity in the silage. In most conventional silo fires, the fire will slowly spread until it reaches the surface of the silage or burns through the wooden silo doors. Once the fire has access to unlimited oxygen, it can develop into a full-blown fire.

Fire may spread to other structures, equipment, or livestock when an unloading door burns through. Burning pieces of the door or hot silage may break out into flames as they fall down the chute and may ignite surrounding materials. In some cases, burning silage has been transferred by auger or conveyed to adjoining buildings where it ignited faster-burning materials.

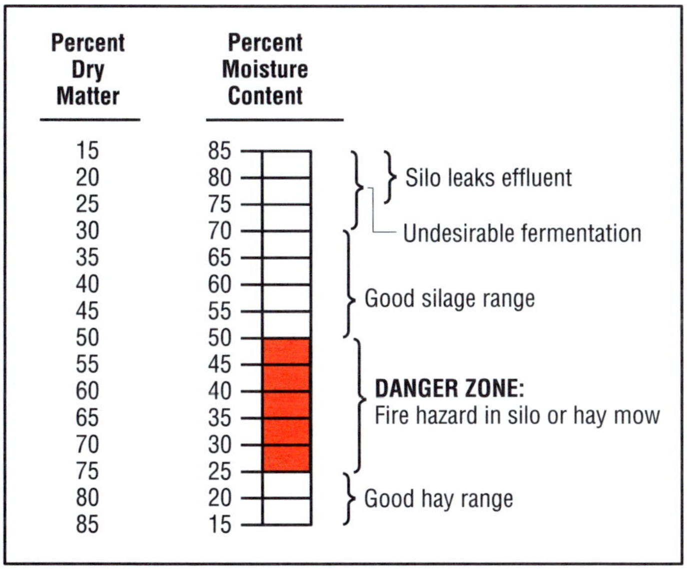

Figure 3.21
Characteristics of stored silage or hay by percentage of moisture and dry matter
Sources: Extinguishing Silo Fires *(NRAES–18),* Silage and Hay Preservation *(NRAES–5)*

Caution: A silo full of silage or a mow full of hay is not worth the life of a firefighter. The procedures recommended here should be applied only during the early stages of the fire. If the fire has already burned deeply into the hay or silage or if excessive temperatures, smoke, or flames are present, the strategy should be one of containment and protection of nearby structures.

Fighting Fires in Conventional Silos

Recent efforts demonstrate that most silo fires identified early enough can be extinguished with only a few firefighters and with minimum risk and relatively light use of an extinguishing agent. Fire-fighting procedures vary with each fire, but general guidelines follow.

Analyze the Fire

Usually there is ample time to analyze the location and extensiveness of the fire at the scene. Occasionally a fire will be well-established or immediately threatening to other buildings. More likely, the silage will glow red with an occasional flame, as in a charcoal fire. For some silo fires, there will be no visible flame or glowing embers, just smoke rising from the beneath the top layer of silage. Fires in conventional silos often originate near the unloading doors where there are air leaks, although a fire can occur at any point where the crop material was not stored at the proper moisture content.

Essential Safety Precautions

1. Before exercising any task inside the silo, determine the location of the silo unloading mechanism. Ask the owner for the location of the unloader, and verify the location by visual inspection. The unloader must then be secured with a chain to the upper support members or lowered completely to rest on top of the silage. Lock out power to avoid accidental start-up of unloading equipment or an electrical shock from a burned or shorted-out power cord.

2. Confined-space entry procedures should be used when fighting silo fires (see sidebar on page 35). Self-contained breathing apparatus (SCBA) is required for all personnel working inside the silo or attached feed rooms or chutes. A lifeline and harness with an appropriate support team are also required. By using overhead support and pulleys, the support team can remain on the ground. Minimal slack should be maintained in

lifelines to minimize any potential falls or entrapment if the surface of the silage should give way.

3. Station an additional observer/rescuer on the silo blower platform or inside the chute to maintain visual contact with the firefighter in the silo.

4. Once inside the silo, do not step directly onto the silage, whether a fire has been confirmed or is only suspected. The fire may create cavities in the silage.

   Lay wooden planks or large pieces of plywood across the silage to provide a work platform, distribute weight over a larger area, and minimize the risk of falling into a cavity the fire may have created. Planks or plywood may be hauled up the chute on some silos, or the plank or plywood may have to be lowered from an opening in the top. (The unloader can also offer a platform to work off.)

Extinguishing Fires

Caution: Use the following procedures in well-ventilated conventional silos only. (For cases of fires in oxygen-limiting silos, see page 50).

When flames are visible in conventional silos, douse fires with water from the silo blower platform or silo chute. Only one firefighter, in full turnout gear, needs to do this. A straight tip nozzle is more effective than a fog nozzle. The water stream from the straight tip nozzle penetrates the pile and better extinguishes fire that has become deep-seated. A ⅜-inch tip is recommended. After the surface fire is extinguished, follow the procedures below for a subsurface fire.

As with any fire in an enclosed area, considerable amounts of gases, smoke, and steam will pour out when the fire in a conventional silo is hit with water. After dousing, remove as many unloading doors and coverings as possible above the top surface of the silage to allow hot gases to escape.

Silo manufacturers do not recommend applying water to the outside of a silo to cool it down. Less stress is placed on the heated silo wall by allowing it to cool down naturally.

If you only suspect a fire or if you can see a burned cavity but no flames, dousing is not necessary. Follow the procedures below for locating and extinguishing a fire.

1. Find the exact fire location with a probe and thermometer (figure 3.22). The most common design for a probe uses lengths of ⅜-inch or ½-inch galvanized pipe with a drive point ventilated with at least four holes. The other end of the probe has a ¼ turn ball valve and sufficient adapters to connect the probe to a 1½-inch fire line.

   Lower a high-temperature thermometer (0–500°F) or other temperature-sensing device on a lightweight wire into the probe to obtain temperature readings.

2. Make several temperature readings, starting near any obvious hot spots. Gradually probe out toward the silo walls at 3-foot intervals. Several hot spots may exist because the fire will follow air pockets to support itself. Make a diagram (such as in figure 3.23) to keep track of hot spots.

   A probe will penetrate easily into a spot that has been charred or burned. It may be difficult to push a probe more than 4–6 feet into packed, finely cut silage. Make every effort to insert the probe to a depth of at least 6 feet.

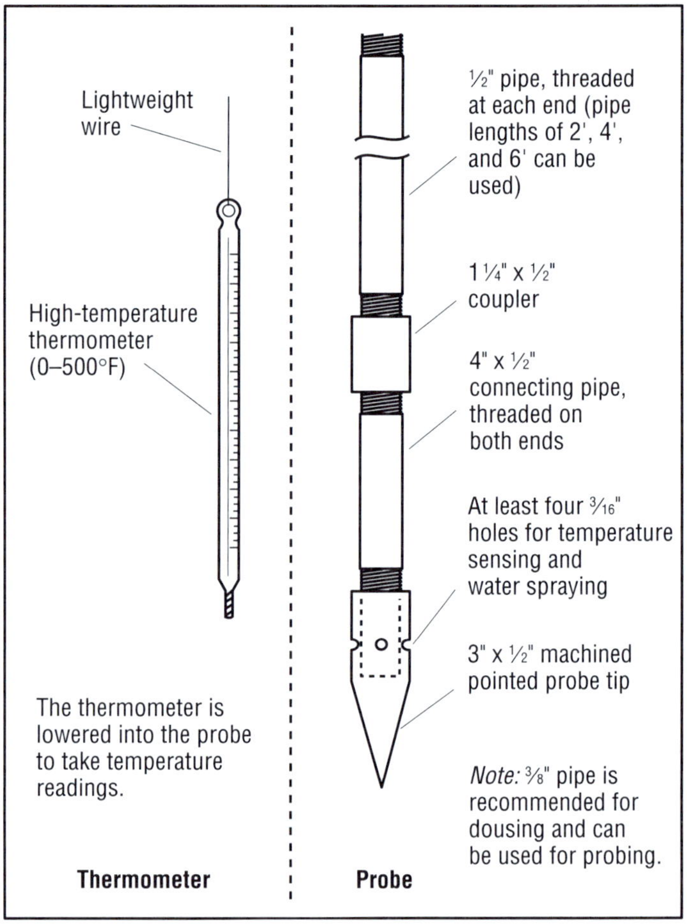

Figure 3.22
Thermometer and probe

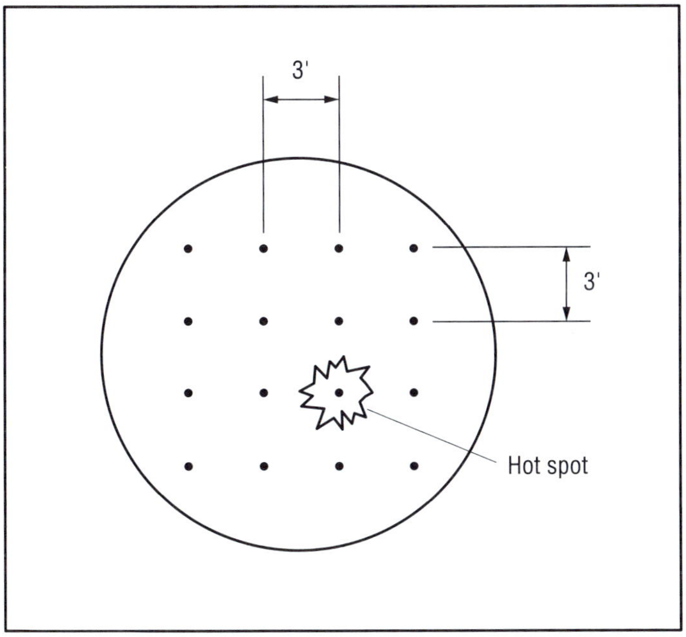

Figure 3.23
Locating hot spots

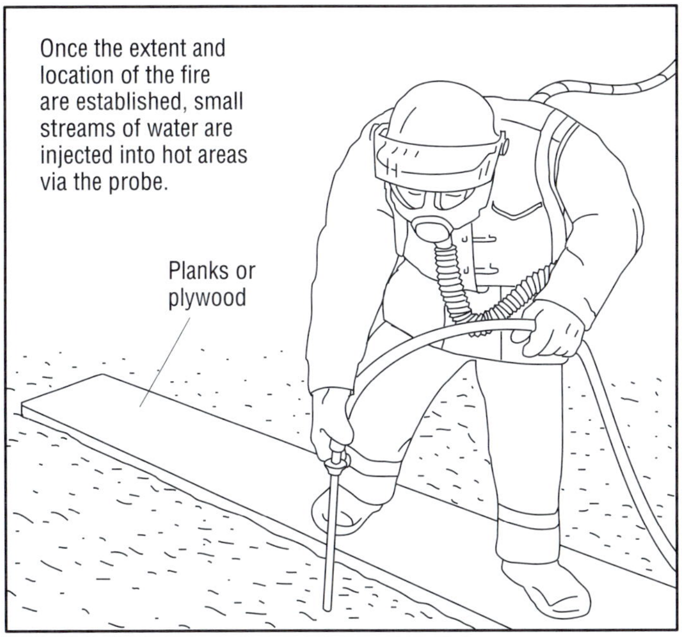

Figure 3.24
Injecting water into hot spots via a probe

3. If temperature readings are near 180°F or higher, the material will eventually burn, and it has lost its value as a feed. Temperature readings at 140–170°F may indicate that silage is heating, or they may indicate that residual heat from a hot spot is moving through the silage. Monitor temperatures every two or three hours at this stage. Temperatures below 140°F indicate no serious heating problems.

4. If the temperature reaches 180°F or higher, action must be taken to cool the silage. This is accomplished by using the probe to inject small streams of water directly into hot areas (figure 3.24). Work slowly and methodically. Leave the injection probe in a single spot for no more than 20–30 seconds before moving it. Injecting water into hot spots is more effective than the traditional method of flooding the top of the silage with water.

   Injecting too much water can be dangerous. Minor explosions have been caused by steam being produced at a rate greater than it can be released from the burned-out cavity.

5. The firefighter operating the injection probe should follow confined-space procedures, including full turn-out gear and SCBA. A second firefighter, also in full gear and SCBA, should be stationed inside the silo chute to help handle the hose and stand by for safety.

6. Because the holes in the probe's perforated tip are only 3⁄16-inch diameter, use the smallest hose line from the fire engine. A garden hose has been used successfully. In many cases, you can inject water from the silo chute, particularly when a hatch door has burned through; but be prepared for considerable amounts of smoke and steam that may blow back.

7. There may be concern about creating a "water-gas reaction" and causing the silo to explode by injecting water into burning silage. This is **not** a realistic expectation in conventional silo fires. The few violent explosions that have been reported from silo fires have all been in oxygen-limiting silos, where the explosive gases have been trapped and could not escape. Even in these cases, it is not certain that the explosions were caused by water-gas reactions.

   The water-gas reaction comes from the reaction of water molecules with burning carbon, producing hydrogen (H_2) and carbon monoxide (CO) gases. This can happen when water is injected into very hot carbon fires. However, the chemical reaction is highly endothermic and causes the temperature of the hot material to drop rapidly. This lowers the rate at which the reaction occurs and then stops the reaction totally. Thus, there would be only very small quantities of gases produced by a water-gas reaction.

Another important point that practically precludes violent explosions from occurring in conventional silos is that there is no containment of gases or air; thus, there can be no explosion. For an explosion in a silo to occur, there have to be containment and consequent buildup of pressure beyond the capacity of the structure. This does **not** happen in conventional silos. However, a few occurrences of minor explosions have been reported. These minor explosions have been caused by rapidly developing steam production that was generated during inappropriate water injection.

Rapid steam production can be a risk when water is injected at very high flow rates and for long durations into areas of heavy combustion. The steam produced under such extreme conditions is produced at a rate greater than its ability to escape from the cavity. Furthermore, the steam has no opportunity to perform its extinguishing function. When injecting water, use minimum flow rates and do not allow water to flow for more than 20–30 seconds in a single spot.

Remove Damaged Silage after a Fire

When the fire has been extinguished, unload damaged silage and deposit it at a safe distance from any buildings for three reasons:

1. Overheated silage loses its nutritional value.

2. The top layers of the wet silage will spoil.

3. Any hot spots missed might reignite.

In some cases, the top unloader may not be usable because of wetness of the silage or an uneven surface. Some of the silage may have to be removed by hand. Follow confined-space procedures (see page 35).

As layers of silage are removed, take additional temperature readings and examine the silage to determine its condition. It is important to unload the silage to a level just below any burned or charred silage. Unloading below this level is necessary only if there are hot spots, fire-damaged silage, or water-saturated silage.

Most silo unloader motors are designed for intermittent operation and will overheat if run continuously. Hot silage may flame up and burn the motor as layers are exposed to the air. If the unloader is used to remove damaged silage, allow the unloader motor to cool off every half hour. Use the cooling time to probe for hot spots.

Fighting Fires in Oxygen-Limiting Silos

Spontaneous combustion in an oxygen-limiting or sealed silo is rare but can occur. Sometimes a slow-charring fire will self-extinguish due to insufficient oxygen. A fire in an oxygen-limiting silo is potentially hazardous, and proper methods or techniques for extinguishing these fires can prevent explosions. With sufficient heat or flames to ignite the confined gases, the only thing preventing an explosion is insufficient oxygen.

On some silos, warning signs are already in place at the base of the silo adjacent to the silo ladder and at the top of the silo near the closest roof hatch door opening. These signs warn firefighters and other emergency personnel of the danger of using water or foam to extinguish an oxygen-limiting silo fire.

Essential Safety Precautions

1. A fire in an oxygen-limiting silo can be extremely hazardous. If rescuers use improper extinguishing techniques, a devastating explosion could result. Contact the dealer immediately for assistance in identifying trained persons who can extinguish the fire.

2. Do nothing that might increase the oxygen level in an oxygen-limiting silo. Do not open the top-hatch cover; opening the hatch might allow enough oxygen to be pulled in to put gases into their explosive range. Even air trapped in water droplets and foam particles can increase the danger.

3. Check the unloading conveyor door to see if it is properly closed and sealed. If it is not, seal it so that oxygen cannot enter through it to feed the fire.

4. Leave an open roof-hatch alone if smoke or steam is coming from the top, or if the silo shakes or rumbles. If the silo is quiet and you haven't seen smoke for several hours, it should be safe to close the top-hatch cover. However, do **not** tie down or latch the cover. If gas pressure subsequently builds beyond the relief capacity of the breather valve, the cover should be free to open to relieve the pressure.

Fire-Fighting Procedures

Caution: Do **not** introduce water or foam into an oxygen-limiting silo in which there is a fire. An explosion could occur. Be sure to follow the safety precautions listed above.

1. Closing all hatches for one to three weeks may be all that is needed to extinguish a fire in an oxygen-limiting structure. Decisions should be made in consulta-

tion with the farmer and the silo manufacturer's service representative. Periodic inspections should be made during this period to observe any progression of the fire, such as smoke or heating up of the silo wall.

2. One method of extinguishing fires that has been used in these structures is for trained firefighters, silo service representatives, or suppliers of industrial gases to inject liquid nitrogen or carbon dioxide to displace oxygen and cool the fire. However, injecting liquid nitrogen or carbon dioxide requires special training to avoid increasing the risk of explosion. The silo dealer can assist in contacting experienced individuals.

Fighting Fires in Hay Mows

Baled hay continues to cure several weeks after storage. The curing process results in several cycles of heating and cooling. If the moisture content in the hay is greater than 25 percent, more heat may be generated than can be dissipated through the hay and into the atmosphere. The chance of spontaneous combustion increases as the hay stack gets larger, the bales get denser, and ventilation around the hay stack decreases.

Fires from freshly cut hay usually occur within two to six weeks after baling. Even hay baled at the proper moisture content or hay in storage for more than a year can spontaneously combust. Humid air, a leaky roof, or storing hay on the ground can increase the moisture content, creating the conditions for spontaneous combustion.

If the heat generated increases the temperature to 175°F, fire is likely to occur. At lower temperatures, increased ventilation can help dissipate heat. At 175°F or above, increased ventilation feeds the fire and should be avoided.

Smoldering hay gives off a strong, pungent odor. This odor is an indication that a fire is occurring.

Essential Safety Precautions

1. Never investigate a hay mow fire alone. The hay may contain burned-out cavities. Moving bales to locate hot spots may result in a burst of flame as the hot bales come in contact with fresh air.

2. Follow confined-space procedures, including full turnout gear with a self-contained breathing apparatus (SCBA), when approaching a fire. Severely charred or burning hay will emit carbon monoxide. Also, the hay could be treated with a chemical preservative. Even though the most common forms of preservatives (propionic acid and ammonia/urea) are not known to produce toxic gases during fires, new preservatives come onto the market regularly. Preservatives may emit toxic gases at elevated temperatures. Check with the farmer to determine if preservatives are used.

3. Heat, smoke, and the physical demands of moving hay bales have lead to injuries, exhaustion, and heart attacks. Firefighters should be monitored and those exhibiting signs of exhaustion should be rotated off the team.

4. If checking for hot hay or extinguishing burning hay requires walking or standing on the hay, the firefighter must use a lifeline and support team as if in a confined-space situation. If this cannot be done, stay off the hay.

Fire-Fighting Procedures in Hay Mows

1. Douse any visible flames with water.

2. Check the temperature of the hay by using the same probe as described in the section on checking silage for hot spots (page 48). Hot spots with temperatures above 175°F indicates that fire pockets are likely.

3. Minimize air movement around the hay to prevent feeding the fire with fresh air.

4. Inject hot spots with water using the same probe and method described in the section on fighting fires in conventional silos (page 48).

5. Move all heated hay to a remote location away from any buildings or flammable material. Heated bales can burst into flame while they are being moved so caution is required. Open up hot bales and douse with water to prevent re-ignition.

Emergencies Involving Manure Storage Sites and Facilities

Generally, farmers use three types of liquid manure storage systems:

1. large holding tanks located directly under the livestock housing area (figure 3.25)
2. open lagoons or ponds located near livestock housing areas (figure 3.26)
3. silo-type, above-ground slurry storages

Livestock waste stored in large quantities presents a number of hazards to people and animals. The most obvious hazard is the possibility of falling into an open storage structure and drowning. If someone drowns, you will probably have to drag the area; the body may not float to the top.

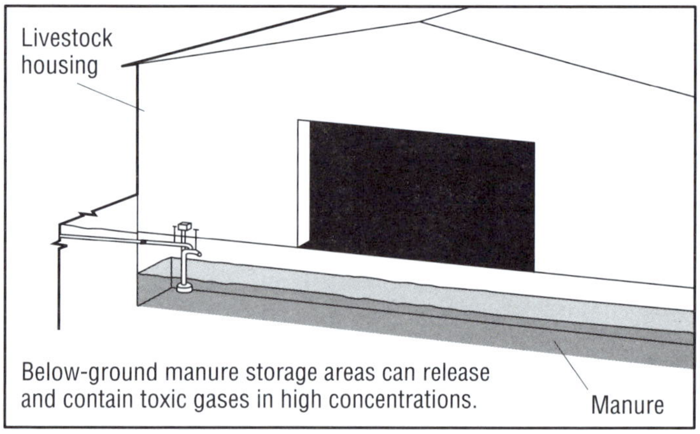

Figure 3.25
Below-ground manure storages

Toxic Gases Released by Manure

The release of various toxic gases caused by bacterial action is a potential hazard from manure stored below the livestock housing area or in any poorly ventilated area. Agitating the liquid, which is often done before the stored manure is pumped, causes more rapid release of gas. The most common gases include ammonia, carbon dioxide, methane, and hydrogen sulfide. See sidebar on page 53 for a brief guide to gases commonly released by stored manure and their effects. Table 3.1 (page 54) offers additional information on gases present in agricultural confined spaces.

Below-Ground Manure Storage Pits

The most hazardous and common type of manure storage rescues involve below-ground manure storage pits, usually in hog or dairy confinement buildings. When a person enters a manure storage pit with a high concentration of hydrogen sulfide gas, the person can die within a few seconds. Such tragedies have been made worse when others enter the pit to rescue the victim and are also overcome. Extreme caution is necessary during rescues of this type.

Toxic gases from the below-ground manure storage pit can impact the air quality in the associated livestock confinement facility. A good indicator of an air quality problem is the condition of the livestock in the building. If the livestock appear to be acting abnormally, assume the presence of toxic gases and follow the procedures below.

Essential Safety Precautions

1. Rescue attempts involving below-ground manure storage pits should follow confined-space procedures (see page 35). Use a self-contained breathing apparatus (SCBA) and a full body harness. The backup crew should also use SCBA (figures 3.27 and 3.28). A mechanical lifting device is preferred for pulling the rescuer from the pit.

2. Never allow the rescuer to disconnect lifelines, switch face masks, or share oxygen with the victim. It takes only a few seconds for a person to be overcome by the toxic gases in a below-ground manure storage pit.

Rescue Procedures

1. Confirm the location of all employees and family members to identify who may be in the manure pit.

2. Ventilate the facility if possible by opening windows and doors or activating a ventilation system.

text continued on page 55

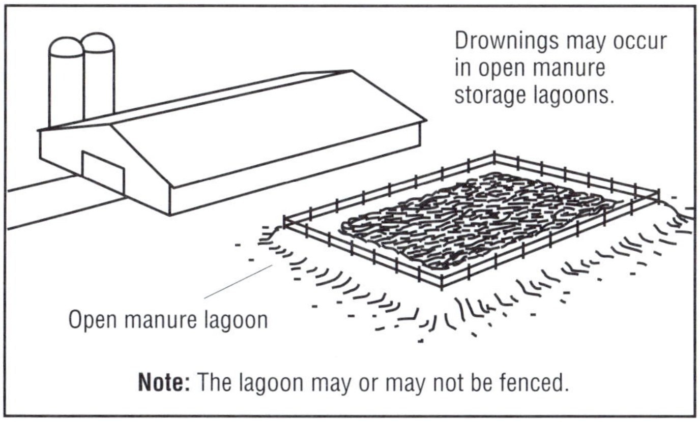

Figure 3.26
Open manure storage lagoon

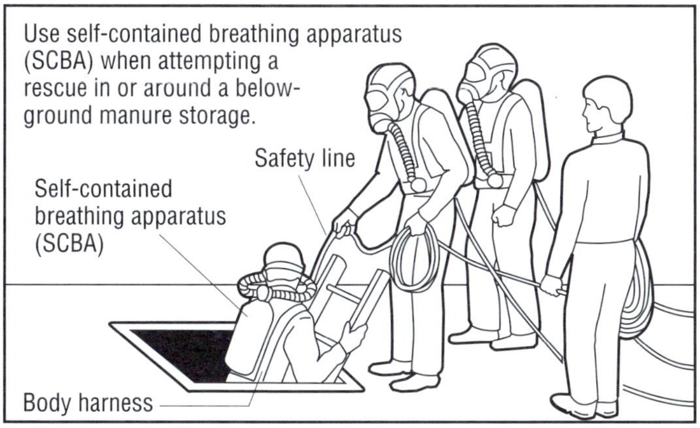

Figure 3.27
Rescuing a victim from a below-ground manure storage

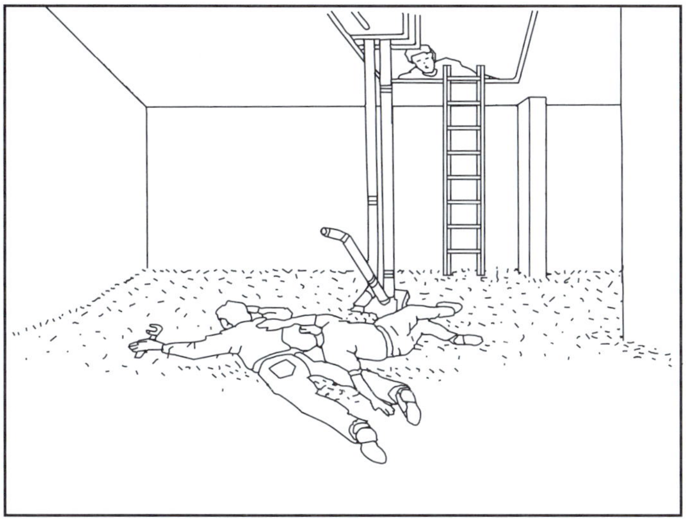

Figure 3.28
This below-ground manure storage has claimed the lives of two persons, the original victim and a would-be rescuer who entered the pit unprotected.

Gases Commonly Released by Manure

Ammonia (NH_3)

Ammonia (NH_3) is a strong alkali that has a pungent odor like that of household ammonia. In small concentrations, ammonia can severely irritate the respiratory system; and in high concentrations, it can be fatal. In cases of ammonia exposure, flush the irritated skin or eyes with large quantities of water and provide oxygen (see page 57).

Carbon Dioxide (CO_2)

Carbon dioxide (CO_2) is an odorless, colorless gas that is slightly heavier than air. If the concentration of carbon dioxide in air exceeds 40,000 parts per million, it can cause death by suffocation.

Methane (CH_4)

Methane (CH_4) is a highly flammable, nontoxic gas. There are reports of fires breaking out in confinement buildings when methane pockets beneath the structure were ignited by welding sparks or open flames. Asphyxiation is also possible in a confined space with a high concentration (5–15 percent) of methane. Methane is extremely difficult to detect without gas-detection instruments because it is odorless; as a precaution, anticipate its presence in all manure storage areas.

Hydrogen Sulfide (H_2S)

Hydrogen sulfide (H_2S) is a very poisonous gas with a strong, detectable "rotten egg" odor. At high concentrations (above 50 parts per million), however, it deadens the sense of smell; so do not assume it is gone just because you cannot smell it after a few minutes. Hydrogen sulfide is the most dangerous by-product of manure decomposition. If the concentration of hydrogen sulfide in air exceeds 300 parts per million, death from respiratory paralysis can occur with little or no warning.

Caution!

Both carbon dioxide and hydrogen sulfide are heavier than air. They usually settle in low areas of the storage facility and remain in high concentrations, even after ventilation.

Table 3.1
Gases present in agricultural confined spaces

Gas	Health Effects		Maximum Exposure Limits IDLH, Immediately Dangerous to Life or Health; REL, Recommended Exposure Limit; STEL, Short-Term Exposure Limit *(note 1)*			Physical Properties			Flammable Properties
	Acute	Long-Term	IDLH (ppm)	REL (ppm)	STEL (ppm)	Relative Density (air = 1) *(note 2)*	Color	Odor	
Ammonia (NH_3)	Irritates eyes, nose, throat; burns skin	Permanent lung damage	300	25	35	0.60 (lighter than air)	None (colorless)	Pungent, suffocating	Should be treated as a flammable gas
Carbon dioxide (CO_2)	Asphyxiant; headache, dizziness, sweating; increased heart rate	None	40,000	5,000	30,000	1.53 (heaver than air)	None (colorless)	None (odorless)	Nonflammable gas
Carbon monoxide (CO)	Asphyxiant; headache, nausea, weakness, dizziness	—	1,200	No standard. 200 "ceiling" *(note 3)*	No standard	0.97 (lighter than air)	None (colorless)	None (odorless)	Flammable gas. Explosive between 12.5% and 7.4% by volume of air mixture. Auto ignites at 1,128°F (609°C)
Hydrogen sulfide (H_2S)	Irritates eyes and respiratory system; coma, convulsions, eye pain, tearing in eyes, dizziness	Eye irritant, lung irritant	100	No standard. 10 "ceiling" (10-minute exposure limit) *(note 3)*	No standard	1.19 (heavier than air)	None (colorless)	Strong odor of rotten eggs	Flammable gas. Flammable between 4% and 44% by volume of air mixture.
Methane (CH_4)	Asphyxiant	—	No standard	No standard	No standard	0.72 (lighter than air)	None (colorless)	None (odorless)	Explosive between 5% and 15% by volume of air mixture. Auto ignites at 999°F (537°C)
Nitric oxide (NO)	Respiratory irritant; irritates eyes, wet skin, nose, throat; drowsiness	—	100	25	No standard	1.04 (heavier than air)	None (colorless)	Strong, pungent	Nonflammable gas, but will accelerate the burning of combustible materials
Nitrogen dioxide (NO_2)	Respiratory irritant; irritates eyes, nose, throat; coughing; chest pain	Permanent lung damage	20	No standard	1	2.62 (heavier than air)	Reddish-brown gas	Strong, pungent	Noncombustible liquid/gas, but will accelerate the burning of combustible materials
Nitrogen tetroxide (N_2O_4)	Respiratory irritant	Permanent lung damage	50	No standard	No standard	1.58 (heavier than air)	Yellow	Strong, pungent	Nonflammable gas, but will support combustion

Adapted from *NIOSH Pocket Guide to Chemical Hazards* (June 1997 edition). Supplemented with information from *Rural Rescue and Emergency Care.*

Note 1: Maximum Exposure Limits are established by the NIOSH (National Institute for Occupational Safety and Health) and are expressed as three distinct values. (a) **IDLH** concentrations are Immediately Dangerous to Life or Health and are expressed in parts per million (ppm). (b) **RELs,** or Recommended Exposure Limits, are time-weighted average (TWA) concentrations for up to a ten-hour workday during a forty-hour workweek. RELs are also expressed in parts per million. (c) **STELs,** or Short-Term Exposure Limits, are fifteen-minute, time-weighted average (TWA) exposures that should not be exceeded at any time during a workday. STELs are also expressed in parts per million.

Note 2: Relative densities of gases indicate how many times a gas is heavier than air at the same temperature. Air is assigned a density of 1. Gases with densities less than 1 are lighter than air; gases with densities greater than 1 are heavier than air.

Note 3: A "ceiling" REL should not be exceeded at any time.

text continued from page 52

3. Only one trained rescuer, with an SCBA and lifeline, should enter the pit with a full support team that can pull the rescuer out if necessary.

4. Lower additional lifelines and air supply to the rescuer using a tag line.

5. Remove the patient to fresh air as quickly as possible. Because inhalation or ingestion of manure can lead to serious medical complications, be sure the patient receives thorough medical attention, even if the patient appears unharmed.

6. The rescue is not over until everyone is out of the pit.

Open Manure Ponds

In open manure storage tanks or ponds, toxic gases are not a danger. However, a drowning hazard exists, and the appearance of the surface crust is deceptive. The crust may look solid and even have weeds growing on it, but it will not always support a person's weight. Always use a small boat or dragline to remove someone who has fallen into an open manure storage (figure 3.29).

Figure 3.29
If the victim manages to stay afloat, the rescue station can provide equipment for a rescue attempt.

Real-Life Scenario

While attempting to unplug a submersible pump at the bottom of a deep manure pit under his hog confinement building, a farmer climbed down into the pit using an extension ladder. His father discovered him a short time later laying at the bottom of the pit with his face in the liquid manure. The father called for help and then proceeded to enter the pit. By the time the local rescue team arrived, a hired employee had also entered the pit and was hanging on the edge of the opening nearly overcome by the gases. Using SCBAs and lifelines, all three men were removed from the pit, and efforts were made to resuscitate each using CPR. Both the father and the son were pronounced dead at the scene, and the hired employee was transported and recovered following medical treatment. An autopsy later revealed that the causes of death were drowning and exposure to hydrogen sulfide gas.

Chapter 4

Farm Chemical Exposures, Fires, and Spills

Exposure to Anhydrous Ammonia

Widely used as a nitrogen fertilizer, anhydrous ammonia (NH_3) is pure ammonia (anhydrous means "without water"). At moderate temperatures and atmospheric pressures, it is a colorless gas, but under pressure it changes into a liquid. Agricultural ammonia is transported, stored, and handled in pressurized tanks in liquid form (figure 4.1).

Escaping anhydrous ammonia forms a white cloud when it combines with moisture in the air. Under conditions of high humidity, the ammonia cloud will remain close to the ground and not break up quickly. Under drier conditions, the ammonia cloud will rise and quickly disperse.

Even mild exposure to ammonia can cause irritation to the eyes, nose, and lungs.

Figure 4.1
Anhydrous ammonia (NH_3) nurse tank and field applicator

Injuries

Because anhydrous ammonia is completely devoid of water, it absorbs water from the skin very quickly. In addition, when combined with water, it forms ammonium hydroxide, a caustic alkaline solution that can cause painful skin burns and, because of its low boiling point, can also burn by freezing. Even mild exposure can cause irritation to the eyes, nose, and lungs. Prolonged exposure can result in suffocation. Small amounts of water near an anhydrous ammonia leak may absorb enough ammonia to cause irritation or blindness if splashed in the eyes. Anhydrous ammonia has a sharp, irritating odor, similar to the smell of the diluted ammonia used for household cleaning. This odor is an excellent warning of the presence of anhydrous ammonia.

Essential Safety Precautions

1. Wear self-contained breathing apparatus (SCBA), eye protection, and rubber gloves when working at the scene of an ammonia spill.

2. Approach the spill site from upwind, if possible.

3. If ammonia fumes could potentially enter a populated area, implement evacuation procedures immediately.

4. Water applied by a fog nozzle can effectively control or disperse ammonia fumes in order to allow access to victims or to shutoff valves.

5. Try to shut off the leak by closing the supply valve nearest the leak. If the valve is not accessible or if adequate protection is not available, the ammonia can be allowed to bleed off and be absorbed by a steady stream of fog from the fog nozzle.

6. An individual cannot remain in an area with a high concentration of ammonia voluntarily due to its irritating nature. A victim who has been overcome by ammonia fumes has probably stopped breathing. Administer basic life support immediately.

Treating the Victim of Anhydrous Ammonia Exposure

Eye Injury

1. Remove contact lenses if the patient is wearing them. Contact lenses trap ammonia behind them, causing more damage. Hold eyelids open and pour water continuously over the eyes (figure 4.2). Irrigation of the eyes should continue until the patient receives medical care. Do **not** let the water run from one eye into the other. Irrigate eyes from the bridge of the nose, and be sure to irrigate under the eyelids. Ten gallons of water will provide a pencil-sized stream for fifteen minutes.

2. If there is a delay before the patient can receive medical attention, provide continuous eye irrigation. If the water supply is limited, juice, cold coffee or tea, or soft drinks can also be used.

Respiratory Injury

1. Irrigate or rinse the nose and throat with water for 15 minutes if the patient is conscious. Be careful not to choke the patient. If the patient is unconscious, place him or her in a side-lying position while irrigating the mouth, nose, or throat.

2. If the patient can swallow, encourage him or her to drink some type of citrus drink, such as lemonade, orange juice, or grapefruit juice. Ammonia is alkaline, and the acidity of the fruit juice will counteract its effects.

3. Place the patient in an area free of ammonia fumes and in a reclining position, but with head and shoulders elevated.

4. Oxygen is useful in treating inhalation exposure to ammonia fumes. Administering 100-percent oxygen at atmospheric pressure is sufficient in most cases.

Skin Injury

1. Do not attempt to remove clothing right away, since a splash of anhydrous ammonia can freeze clothing to skin almost instantly. While still clothed, the patient should get into water or stand under a shower (figure 4.3). Remove or cut off clothes after they have thawed.

2. Flood the affected area with water continuously until the patient receives medical care. Allow the burns to

Even if only a small amount of ammonia enters the eyes, they should be flushed with water continuously until the patient is delivered to medical care. The entire eye surface and inner eyelid must be thoroughly and continuously irrigated.

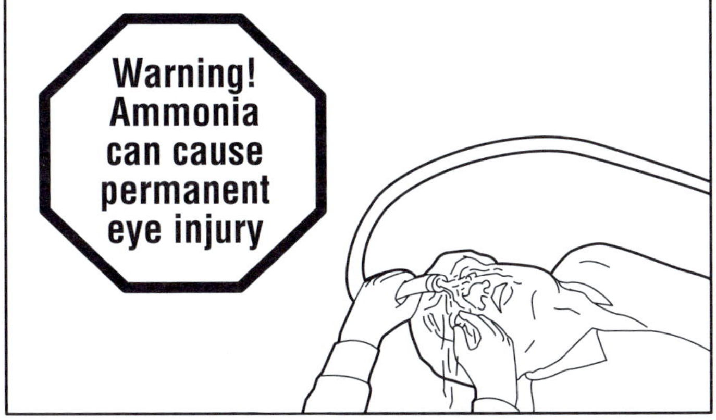

Figure 4.2
Eye irrigation for patients exposed to ammonia

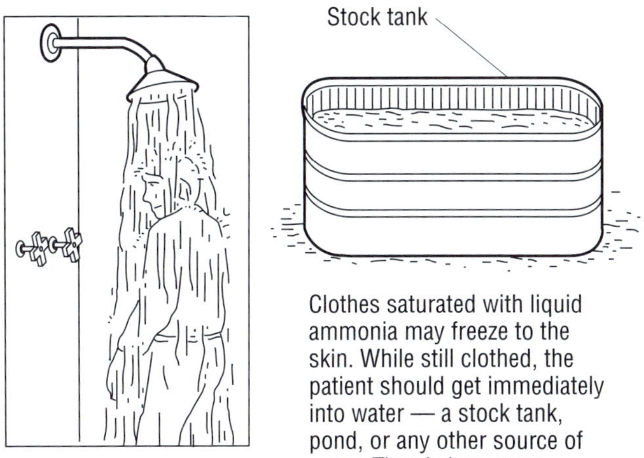

Clothes saturated with liquid ammonia may freeze to the skin. While still clothed, the patient should get immediately into water — a stock tank, pond, or any other source of water. Time is important. Remove clothes after they thaw and can be freely removed.

Figure 4.3
Saturate clothing exposed to anhydrous ammonia with water before removing

remain open to the air.

3. Do not apply salves or ointments to the burns. Be sure to tell medical personnel that the patient's burn was caused by ammonia, so that they will not apply salves or ointments.

4. Watch for signs of dehydration in a patient who has had large areas of his or her body exposed to anhydrous ammonia.

Real-Life Scenario

A farmer was dismantling a plugged regulator on an anhydrous ammonia applicator in order to clean it. In the process, he was suddenly sprayed on the face and body with high-pressure ammonia. Fortunately, the incident occurred in the farmyard, and his wife rushed to him and began flushing him with water from the emergency five-gallon water supply mounted on the ammonia tank. The farmer continued flushing while his wife rushed to the house to call the local emergency medical unit.

Once on the scene, emergency medical technicians continued flushing his face and body with available tap water as they removed his contaminated clothing. They gave special attention to the farmer's eyes to ensure that contact lenses were not being worn and that the eyes were thoroughly irrigated under the eyelids. They continued flushing the eyes with sterile water on the way to the hospital.

Agricultural Chemical Fires

Pesticides (including herbicides, insecticides, and fungicides); fertilizers (including ammonium nitrate, urea, and potash); sanitizers (often calcium hypochlorite); and acids (such as dairy pipeline cleaners) are commonly stored on farms. By themselves, agricultural chemicals do not cause fires; but some are highly flammable or even explosive in their chemical composition. Other chemicals produce highly toxic smoke and vapors when burned or heated (figure 4.4). It is extremely important to know the hazards associated with agricultural chemicals and to take protective measures when fighting fires where they are present.

Fire or rescue units or departments should have readily available a notebook or special file including material safety data sheets (MSDSs) for common agricultural chemicals used in their area. (Businesses selling the chemicals should be able to provide MSDSs.) These sheets have important information on chemical toxicity, fire hazards, treatment for exposures, procedures for dealing with spills, and phone numbers for manufacturers.

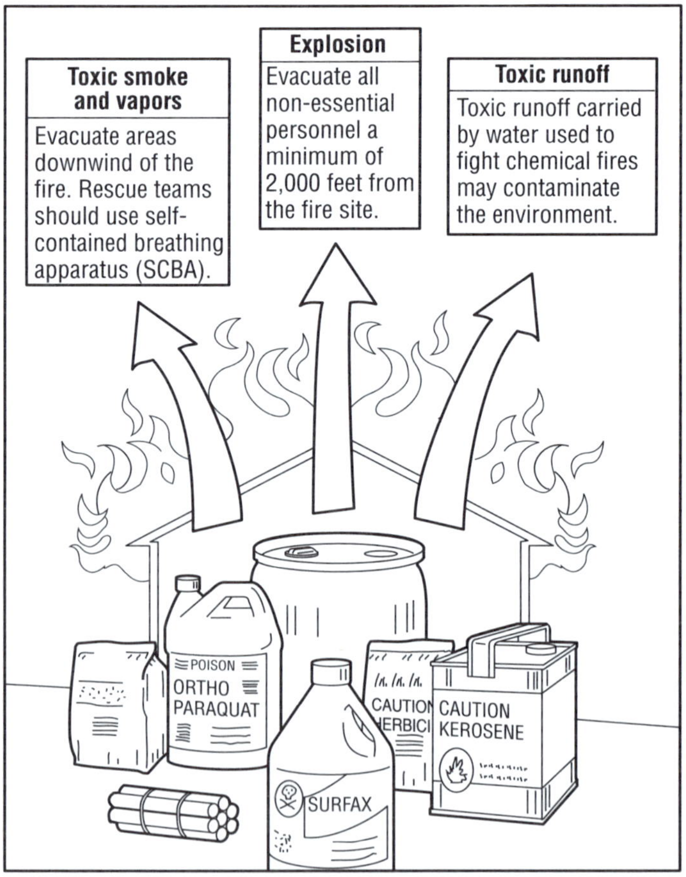

Figure 4.4
Types of farm chemical emergencies

Recommended Procedures for Fighting Chemical Fires

People to Contact Immediately

1. Contact the owner or operator of the facilities involved.

Find out from the owner what the contents of the building on fire are and whether any toxic or explosive chemicals may be present.

2. Notify your local hazardous materials response team (HAZMAT) through the emergency dispatcher.

3. Alert the hospital and poison control center to the possibility that victims were exposed to toxic chemicals and/or their fumes. If possible, tell them the specific chemicals involved.

4. Notify state and local public health officials. Post warning signs and rope off the burned-out area and runoff areas to help prevent unauthorized entry. Consult local HAZMAT teams as well as the manufacturers of the chemicals for help in dealing with these emergencies. CHEMTREC (1-800-424-9300) may provide additional advice (figure 4.5). The state environmental protection agency should also be contacted.

Develop a Fire Response Plan

All owners of chemical-storage facilities, including large farm operations, should be encouraged to file an emergency response plan with the local fire department and police. An agreement with the owner, fire department, and environmental protection agency should be developed to determine the action to be taken if a fire starts.

Let It Burn?

The incident commander in consultation with HAZMAT team members, state and local authorities, and the owner may decide that the best course of action will be to let the building and its contents burn. This may be the best course of action in situations where, if water is applied to a fire where agricultural chemicals are present, incomplete combustion of chemicals will release toxic compounds into the air, or where contaminated runoff may pollute nearby streams and ponds.

If this strategy is used, protecting adjacent structures will be vital. This protection needs to be done from an upwind side. Firefighters should evacuate people and animals downwind as necessary. Ventilating the fire for more complete combustion reduces dangerous toxic smoke. Ventilation should only be done by experienced firefighters wearing self-contained breathing apparatus (SCBA).

Essential Safety Precautions

All firefighters and rescue personnel in the fire area must wear appropriate personal protection (figure 4.6).

Fire-Fighting Procedures for Chemical Fires

1. Attack a chemical fire from upwind or crosswind (figure 4.7, page 60).

2. Use soft streams of water to avoid tearing open paper containers.

3. To keep runoff and potential pollution to a minimum, avoid using large volumes of water. If necessary, dike the area to contain contaminated water (figure 4.7, page 60).

Figure 4.5
Contacting help for a chemical fire emergency

Figure 4.6
Protective clothing and respiratory protection are essential for chemical fire rescues

Treating the Patient Exposed to Toxic Smoke or Fumes

1. When firefighters or rescuers are exposed to fumes or smoke without adequate respiratory protection, immediately relieve them from duty and check for possible poisoning. Poisoning symptoms include headache, giddiness, blurred vision, coughing, tightness of chest, salivation, twitching, or pupils failing to contract when eyes are exposed to bright light.

2. Follow proper decontamination procedures as established by your local HAZMAT team or, in an emergency, as summarized in the following section. Treat anyone exposed to organophosphates (common in pesticides) immediately. Do not wait for symptoms to occur. Organophosphate poisoning symptoms may be mistaken for heat exhaustion or smoke inhalation. Anyone exposed to organophosphates should receive cholinesterase tests. Obtain a label that describes the chemical to help guide those treating the patient.

3. Symptoms of chemical exposure may be delayed up to twelve hours; some symptoms may be delayed several days.

4. Clothing may retain toxic fumes and endanger personnel when they remove their self-contained breathing apparatus. Be sure to carefully follow decontamination procedures.

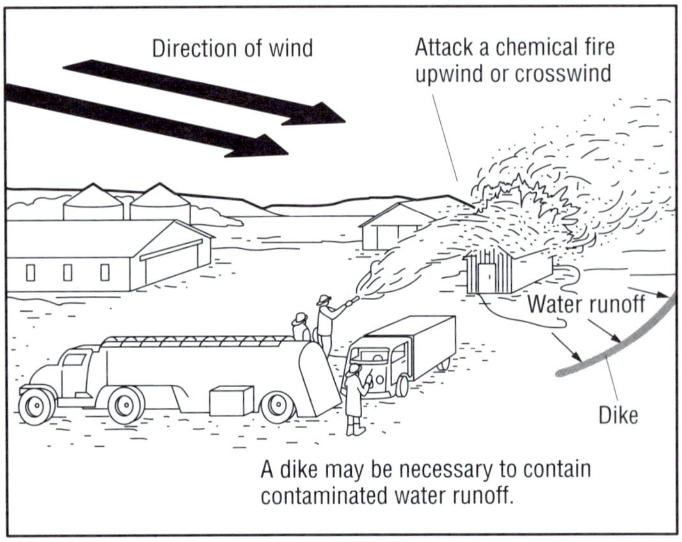

Figure 4.7
Fighting a chemical fire

Procedures for Decontamination after a Fire

Before Decontamination

1. Notify state and local public health officials. Post warning signs and rope off the burned-out area and runoff areas to help prevent unauthorized entry. Contact local hazardous materials response teams and chemical manufacturers for expert advice in decontamination procedures. CHEMTREC (1-800-424-9300) may provide additional advice. The state environmental protection agency should also be contacted.

2. Use an expert decontamination team, if available, or a hazardous-materials response team. If such a team is not available, a person familiar with pesticides should direct the cleanup operation and use the procedures below.

3. Be sure that all personnel involved in cleanup understand the toxic nature of the debris, are properly clothed, and have adequate respiratory protection.

Neutralizing and Disposing of Toxic Chemicals

1. Neutralize and absorb toxic chemicals by covering them with twice their volume in lime, special clay absorptive material, or soda ash; then dampen the area slightly. Spray out-of-reach surfaces, such as standing walls or joists, with a solution of 50 pounds of lime per 100 gallons of water.

2. After neutralizing toxic chemicals, close off the area overnight, being careful not to track pesticides out of the site.

3. Neutralize runoff water by adding large quantities of soda ash. Pump standing water immediately into appropriate disposal containers for disposal in an approved dumping area.

4. Absorb excess liquid with lime or approved absorbent material. Do not wash any material into a waterway or sewer system without official authorization from public health officials.

5. Use mechanized loaders, dump trucks, and other handling equipment in order to minimize human contact with the contaminated materials. Use equipment that will minimize the amount of dust raised.

6. The next day, carry debris to an approved dumping area in metal-bodied dump trucks without holes or leaks, or in tightly sealed containers.

7. Cover the truckload with a disposable cover if dust is a problem. To prevent spills, do not overload trucks.

Decontaminating People

Anyone involved in a chemical fire or decontamination should immediately remove any traces of toxic chemicals by bathing with large amounts of soap and water. Be particularly careful to clean under fingernails. If leather-soled shoes were worn, check them for absorbed pesticides. Properly dispose of all contaminated footwear and clothing. Wash respirators and all protective clothing without delay. Be careful not to contaminate the inside of the respirator with chemicals deposited on the outside. Install new cartridges in respirators. Watch for symptoms of pesticide poisoning, which may not appear until several hours or even days after exposure.

Real-Life Scenario

Late one evening, a farmer noticed a small blaze through the window of his machinery storage shed. He quickly called the nearby volunteer fire department, and they arrived on the scene within minutes. By the time firefighters located a water source and set up their equipment, much of the storage building was in flames. The fire department had only one self-contained breathing apparatus; and, since the farmer had indicated that only machinery was being stored in the building, none of the firefighters used respiratory protection while extinguishing the blaze.

Shortly after the fire was put out, several of the firefighters began to vomit and some exhibited other symptoms of some form of poisoning. The emergency medical personnel at the scene administered oxygen and transported the sick firefighters to the hospital for further treatment. The examining physician noted that the symptoms exceeded those normally observed in smoke-inhalation cases.

Investigations later revealed that an eight-row corn planter, stored in the building at the time of the fire, was loaded with approximately 80 pounds of granular insecticide and over 600 pounds of dry fertilizer. When these agricultural chemicals were consumed in the fire, they produced a combination of extremely toxic smoke and gases. For the unknowing firefighters, the exposure to these toxic by-products of the fire could have proven fatal. Fortunately, none of the firefighters were permanently affected.

Pesticide Exposure and Spills

Essential Safety Precautions

1. Until proven otherwise, assume any spilled pesticide is highly toxic, flammable, and explosive.

2. At the spill scene, wear disposable protective coveralls and unlined nitrile or neoprene gloves or composite gloves and boots. Some fumigants, such as methyl bromide, are readily absorbed by neoprene, so check the label to see if fumigants are involved. Protect eyes and face with face shields or goggles. If vapors or odors are evident, use a self-contained breathing apparatus (SCBA).

Rescue Procedures — Pesticide Exposure and Spills

Get Vital Information

1. Immediately read the label on the pesticide container. It will give the toxicity of the product and the first aid to be administered. Take the label with the patient to the hospital emergency room.

2. Call the manufacturer of the chemical in question, CHEMTREC (1-800-424-9300), or the local poison control center for further treatment information.

Treat the Victim

1. Remove contaminated clothing from the victim if injuries permit.

2. Use soap or detergent and water to wash the pesticide from the skin. This might best be done at the spill scene if water is available, in order to prevent contamination of rescue vehicles and medical facilities.

Isolate and Confine Spill

Caution: Do **not** hose down a chemical spill; this action spreads the chemical.

1. Keep people away from spilled chemicals by roping off the area. Do not leave the spill scene without stationing someone to remain at the scene to warn of the danger.

2. Confine the spill. If spilled material starts to spread, dike it up with sand or soil. Use an absorbent material (such as soil, sawdust, or a specially designed product) to soak up the spill. Shovel all contaminated material into leakproof containers and dispose of it at an approved site.

3. If a major pesticide spill occurs, keep people away, give first aid if needed, and confine the spill. Then call the local HAZMAT team, the manufacturer of the chemical in question, or CHEMTREC (1-800-424-9300).

Contact Authorities

If bodies of water or waterways are contaminated, notify regional, state, or federal water quality or water pollution authorities; state health officials; and the state fish and wildlife agency. Your state may require that chemical spills be reported to the state environmental protection agency or other designated agency, such as state departments for agricultural chemicals.

Chapter 5

Farm Animal Incidents

Among the most common causes of farm-related injuries, especially on beef, dairy, and horse operations, are livestock incidents. Even though horses, pigs, cattle, and sheep are considered domesticated, all of these animals can seriously injure or kill people, including experienced handlers. Some exotic livestock, such as bison and ostriches, have also been involved in serious injuries to their owners.

A rescuer inexperienced with animals is no match for a frightened, injured, sick, or protective animal, such as a female with newborn offspring. All animals at the rescue scene should be approached with extreme caution. When attempting to move animals to secure the scene of an incident, take into account the information on animal flight zones given in the following rescue procedures. Although in some cases animals can safely be moved and contained, in other cases it may be necessary to shoot or otherwise destroy an animal to reach an injured person or to protect rescuers.

Injuries

Injuries associated with livestock incidents can be diverse. They include bites from dogs, horses, and pigs; kicks from horses and cattle; crushing and internal injuries from being mauled, trampled, or stepped on by horses or cattle; lacerations and puncture wounds from being gored by bulls and other animals with horns; and multiple injuries from being dragged by horses. There are even cases of pigs causing extensive tissue loss or death when they have attacked people who have fallen into a pen. A sow with young piglets can be extremely aggressive.

Rescue Procedures — Animal Incidents

Essential Safety Precautions

1. Approach any unfamiliar or obviously frightened animal with caution. If time allows, locate the owner, or have the local animal control officer come to the scene to remove or contain the animal.

2. Never chase an animal. Pursuit by a person or vehicle will further excite the animal and may lead to additional attacks on the victim or rescuers.

3. When dealing with an aggressive animal, it may be necessary to shoot or otherwise destroy the animal in order to rescue the victim and safeguard rescuers.

4. Avoid the blind spot at the rear of an animal (figure 5.1, page 64). If you approach an animal from its blind spot, the animal may kick you.

Freeing the Victim

1. If the injured individual is still in close proximity to the animal that caused the injury, the animal may be possessive of its prey and extremely hostile if approached. If the animal cannot be moved into a secure area without risk to the rescuers, it may have to be **destroyed in order to save the victim.** Angry bulls have been known to rip through wire and board fences to attack rescuers and prevent them from rescuing the bulls' victims.

2. In cases of injuries in the presence of pigs, the patient

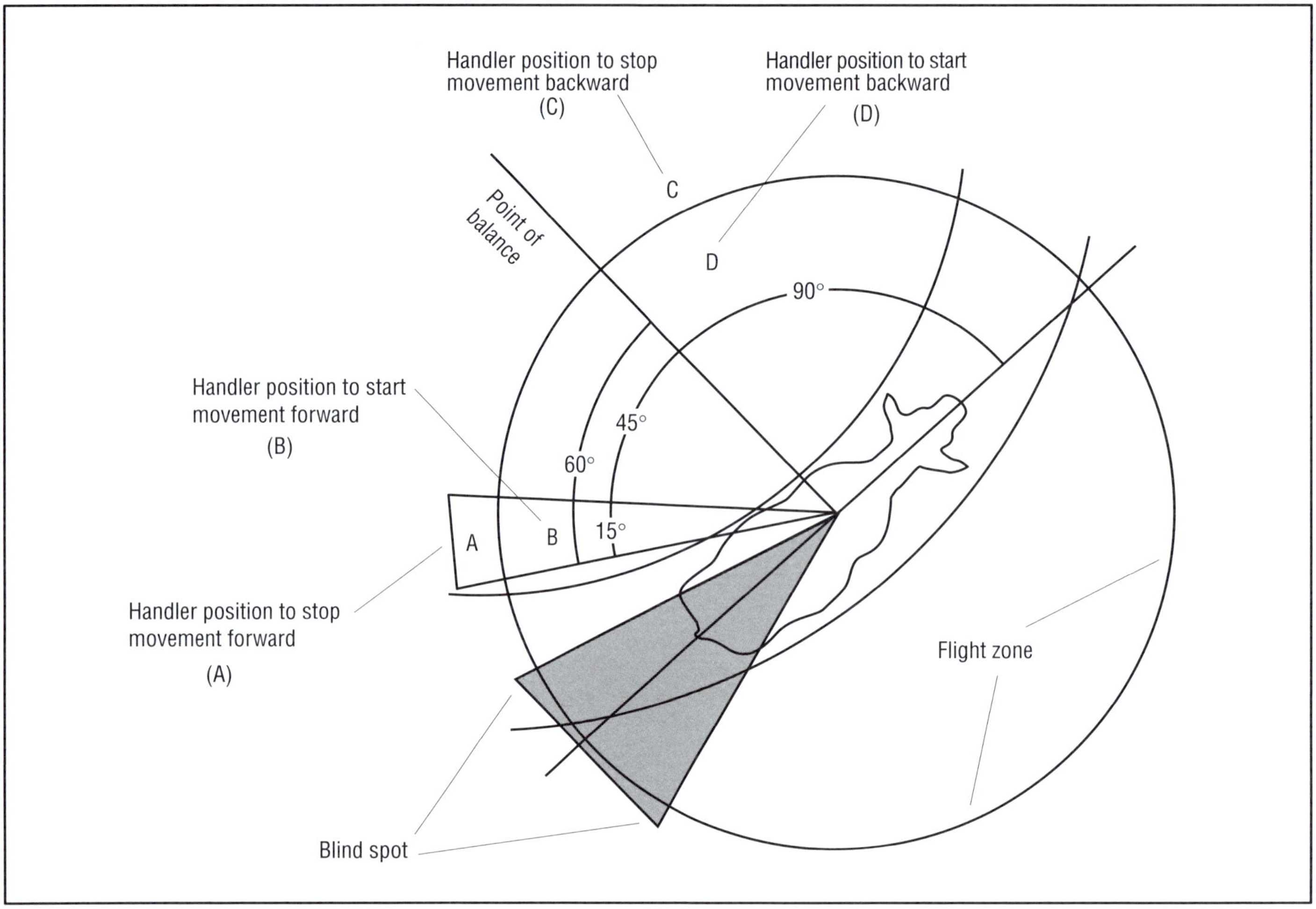

Figure 5.1
Flight zone and animal-handling strategies
Source: Dairy Reference Manual *(NRAES–63)*

should be removed from the area as soon as possible. Pigs will eat human flesh, and the smell of blood can agitate them.

3. Cattle, bison, and sheep are herd animals who attempt to maintain visual contact with one another. A single lone animal that is separated from its herd can become very excited and may be more dangerous than a group of animals. To help calm a lone animal that is charging or agitated, move it into a pen or pasture with the remainder of the herd, if possible.

4. If an animal rears up, do not touch, push, or hit it. The animal is rearing in an attempt to increase the distance between itself and you, and it is likely to become more agitated if touched. The animal will usually calm down if it is left alone.

5. The concept of animals' personal space, or "flight zone" (figure 5.1), should be understood in order to interact more safely with animals. Normally when a person enters an animal's flight zone, the animal will attempt to move away, and it will become upset if unable to move away. Note, however, that an animal that flees is motivated by fear. An animal that is behaving aggressively may not be afraid and so may not behave in ways related to its flight zone.

   The flight zone is a "circle" surrounding the animal, with the size of the flight zone depending on the animal's tameness. For range cattle, the radius of the circular area of the flight zone may be as much as 160 feet; for very tame dairy cattle, the flight zone may be nearly zero and the animal will allow people to touch it.

6. If help from an experienced animal handler is not available, take advantage of the animal's flight zone in order to move the animal away from the scene. To avoid distressing the animal, work on the edge of the flight

zone; do not move too far into the flight zone and do not enter the blind spot behind the animal. The edge of an animal's flight zone can be determined by walking slowly toward the animal from the front and watching for signs of excitement or movement. If you have entered too deeply into the flight zone, the animal will bolt and run away or, if escape is not possible, turn back and run toward you. At the first sign that an animal is turning around toward you, back up and retreat from inside the flight zone.

7. Figure 5.1 illustrates the correct position for moving livestock. Generally, the animal will move in the opposite direction from which he or she has been approached. To make an animal move away, the person handling the animal slowly moves slightly into the flight zone. The handler backs up to outside the flight zone if he or she wants the animal to stop moving. When a person approaches from the front, the animal will turn right if the person moves left. The animal will turn left if approached from the right front.

   To make an animal go forward, work on the edge of the flight zone, advancing slightly inside the flight zone (to start movement) or falling back slightly outside the flight zone (to stop movement) (figure 5.1). To cause the animal to walk forward, stand behind the point of balance (behind the shoulder) slightly inside the flight zone (point B in figure 5.1). To stop the animal from walking forward, back out of the flight zone (point A in figure 5.1). To make the animal move backward, move in front of the point of balance, slightly inside the flight zone (point D in figure 5.1). To stop the animal from moving backward, back up outside of the flight zone (point C in figure 5.1).

8. Very tame animals, which have no flight zone, cannot be moved as outlined above, but must be lead.

9. If a number of animals must be moved from the scene, never chase them or yell at them, as this resembles the behavior of a predator. You can move all the animals by concentrating on moving the lead animal, whom the others will naturally follow. Observe the correct position in relation to all the animals' flight zones and calmly herd them at a walk. Avoid rushing, isolating, or otherwise stressing any one animal, because panic behavior can spread to the rest of the group.

10. Pigs, sheep, and cattle have a tendency to move from a dimly illuminated area to a more brightly lit area, provided that the light does not glare into their eyes.

11. Avoid subjecting animals to moving or flapping objects that may cause them to balk.

12. Livestock have color perception and may balk at a sudden change in color or texture.

13. Avoid sudden loud or unusual noises, which can be very stressful to livestock. However, small amounts of noise can be used to move some animals. Cattle and sheep will move away from a rustling piece of plastic, for example.

14. All livestock facilities are heavily contaminated with bacteria and viruses. Open wounds allow these agents to enter the body, where they may cause tetanus or secondary infections. Alert medical personnel to the conditions at the scene. Rescuers who experience an opening of the skin due to an abrasion or laceration should also be monitored.

15. Check the vaccination history of anyone who has sustained an abrasion or laceration in a livestock facility. If a person's tetanus vaccination is out of date, he or she should be revaccinated immediately.

16. Any animal who has bitten a person should be tested for rabies and other serious communicable illnesses.

Real-Life Scenario

An experienced cattle owner was checking on his cow herd in a remote pasture in which he also kept his herd bull. He had owned the bull for several years, and no one could recall that the farmer had previously had any problems with it. When the farmer did not return for dinner, his son went looking for him. In the pasture the son found his father's badly mangled body with the bull standing over it. When the son attempted to approach his father's body, the bull would grasp the man's belt in its teeth and drag the body away.

The son returned home and called for assistance. The bull had to be destroyed in order for rescuers to reach the body.

Appendix A

Rescue Tool Inventory Guidelines

Having the right rescue tools and knowing how to use them will enable rural rescue units to respond more effectively and safely to most farm-related emergencies. The following lists of tools have been developed by a team of professionals familiar with farm rescues. They have found these tools essential or extremely useful in carrying out the full range of rescues that a rural rescue unit may encounter. Many units will already have many of the items listed.

Rescuers might also consider using tools and equipment that may be readily available at the rescue site. Most farm operations have a wide variety of tools that could be used if necessary. Be cautious, however, and be sure to assess the condition of such equipment. Ropes, chains, jacks, and other lifting equipment may not be in good condition; or they may not be rated to meet the demands of rescue use.

Tools

Power Spreader Unit (Amkus™, Hurst™, and Lukas™ Tools)

	Quantity and Item	
____	1	Spreader unit
____	1	Cutter unit
____	1	Power unit
____	1	Hydraulic hose, 32-foot
____	24	Bottles of oil additive, 2 ounces each
____	1	UL-listed gasoline safety can, 2-gallon, with hose/funnel attachment
____	1	Spare spark plug and necessary wrench
____	1	Quart hydraulic fluid
____	1	Spare power plant recoiler unit with rope
____	2	Clevis link/shackle units with pins
____	2	Case-hardened steel chains (½" x 12') with grab hooks at each end
____	1	Pair, spreader-unit training tips with pins
____	1	Spare pin for spreader unit
____	1	Spare pin for shackle unit

Recommendation: Store the power plant, spreader unit, cutter unit, and hydraulic hoses pre-connected.

Rescue Air Bag System

	Quantity and Item	
____	1	Air bag unit with complete accessory components
____	1	Nylon strap with ratchet, 7,000-pound-capacity, 1" x 120'
____	1	Protective cap for bag nipples
____	2	Open-weave steel mesh base plates, 36" x 36", **or** plywood sheet base plates, 36" x 36" x ¾"
____	10	Compressed air cylinders, 45-cubic-foot (2,216 PSI) or equivalent air supply source

Recommendation: Store the air source, regulator, control unit, and all hoses pre-connected.

Remote Controlled Hydraulic Jack Unit (Porto-Power™)

Note: A 10-ton capacity unit is preferred over a 4-ton unit

Quantity and Item

____	2	Hydraulic pumps
____	2	Hydraulic rams, 10-inch (10 tons)
____	1 ea	5", 10", 18", and 30" extension tubes
____	2	Double male adapters
____	2	Double female adapters
____	5	Lock-on connections
____	9	Locking pins
____	1	Slip lock adjustable extension
____	2	Serrated heads
____	2	Vee heads, 90°
____	2	Wedge heads
____	2	Rubber flex heads
____	2	Flex bases
____	1	Plunger toe
____	1	Cylinder toe
____	1	Chain pull plate
____	2	Chains, 8' x ⅜" diameter, with grab hook on one end only
____	1	Chain shortener, 10" x ⅜" diameter
____	1	Clamp head (some units use a 2-piece head/toe attachment)

Cribbing Blocks

Note: Use cribbing blocks of such hardwoods as oak or maple, which resist splintering and splitting. Most cribbing blocks can be stored at the firehouse.

Quantity and Item

____	18	Hardwood blocks, 2" x 4" x 18"
____	10	Hardwood blocks, 4" x 4" x 48"
____	40	Hardwood blocks, 4" x 4" x 18"
____	10	Wedge blocks, 4" x 18"; wedges cut from corner to corner (figure A.1)

Forcible Entry Tools

Quantity and Item

____	1	Flathead ax, 6-pound, with 36-inch fiberglass handle
____	1	Pick head, 6-pound, with 36-inch fiberglass handle
____	2	Halligan-type entry bars
____	1	Pry-ax-type entry bar
____	1	Sledgehammer, 8-pound
____	2	Baling hooks (hay hooks)
____	1	Pry bar, 66-inch
____	1	Bolt-cutter tool, 36-inch

Cutting Tools

Quantity and Item

____	4	Hacksaws, 12", one-piece frame type with shatterproof blades
____	12	Spare shatterproof hacksaw blades
____	2	Squirt cans with lightweight cutting oil
____	1	Pair of tin snips
____	1	Air impact wrench and sockets
____	1	Air chisel, 300 PSI unit preferable
____	2	Air chisel double-panel cutting chisel bits
____	2	Air chisel flat-chisel cutting bits, 8-inch to 11-inch
____	4	Air cylinders, 45-cubic-feet (2,216 PSI) or equivalent

"Cutting Tools" list continued on next page

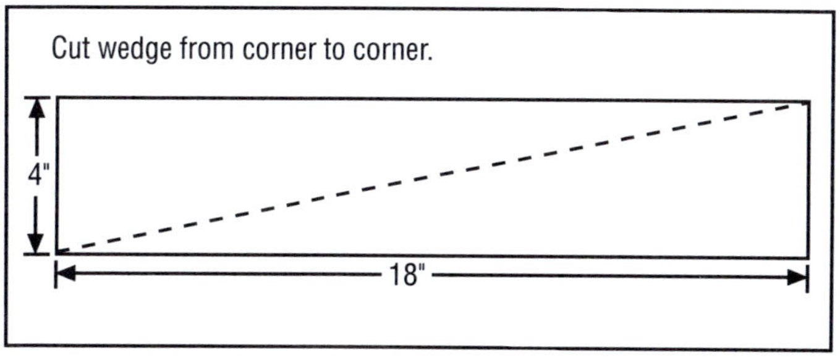

Figure A.1
Cutting wedge blocks

"Cutting Tools" list continued from previous page

- ____ 1 Electric or pneumatic reciprocating saw
- ____ 12 Reciprocating saw blades, shatterproof type
- ____ 1 Can-opener type metal-cutting tool
- ____ 2 Pairs of industrial-quality earmuff-type hearing protection
- ____ 1 Chain saw, 24-inch with support accessories
- ____ 1 Oxyacetylene cutting torch outfit with accessories
- ____ 1 Rotary power saw with support accessories including blades for concrete and metal
- ____ 1 Exothermic cutting torch with rods

Pulling Tools

Quantity and Item

- ____ 2 Hand winch tools (Come-along™), 2-ton capacity
- ____ 2 Alloy chains with hooks, 9⁄32" or 3⁄8" x 12', grade eight alloy
- ____ 2 Alloy chains with hooks, 9⁄32" or 3⁄8" x 15', grade eight alloy
- ____ 2 Alloy chains with hooks, 9⁄32" or 3⁄8" x 10', grade eight alloy

Ropes and Rigging

Quantity and Item

- ____ 2 Nylon kermantle lifelines, ½" or 5⁄8" x 300'
- ____ 2 Nylon kermantle lifelines, ½" or 5⁄8" x 150'
- ____ 2 Nylon kermantle accessory lines, 3⁄8" x 100'
- ____ 1 Polypropylene line, ½" x 100'
- ____ 15 Rated carabiners
- ____ 2 Full-body rescue harnesses with safety lines
- ____ 3 Rated pulley snatch block tools
- ____ Assorted lengths of 1-inch nylon webbing straps

Hand Tools

Quantity and Item

- ____ 2 Utility knives
- ____ 1 Adjustable wrench, 12-inch
- ____ 1 Adjustable wrench, 8-inch
- ____ 1 Set of ½-inch drive socket wrenches, standard and metric, including a 15-inch flex handle
- ____ 1 Set of open-end wrenches, standard and metric
- ____ 1 Flat-head screwdriver, 3⁄16" x 6"
- ____ 1 Flat-head screwdriver, 3⁄8" x 8"
- ____ 1 Phillips head screwdriver, no. 2
- ____ 1 Channel lock pliers, 8-inch
- ____ 1 Set of vise grips, 10-inch with wire-cutter feature
- ____ 2 Mallets, 3-pound
- ____ 2 Pry bars, 12-inch
- ____ 2 Automatic spring-loaded center punches
- ____ 1 Solid-type center punch
- ____ 2 Cold chisels, 1" x 12"
- ____ 1 Finger ring cutter tool
- ____ 2 Ripping-type crowbars, 24-inch
- ____ 1 Tape measure, 1" x 12'
- ____ 1 Chain binder, 3⁄8"
- ____ 2 Stainless steel scissors with serrated blades
- ____ 1 Carpenter-type leather tool pouch with waist belt

Safety/Control Tools

Quantity and Item

- ____ 2 Packages of golf tees for temporary leak stoppage
- ____ 12 Road flares, 30 minutes each
- ____ 1 Container of oil-absorbent material, 5-gallon (stored in quantity at firehouse)

____	100	Square industrial-type oil-absorbent pads, ⅜" x 18"
____	8	PVC fluorescent orange traffic cones: 18-, 24-, or 30-inch-high
____	2	Rolls of duct tape, 3" x 60'
____	2	Large fire safety blankets, 100-percent wool
____	1	Salvage cover, 12' x 20'
____	1	Push broom, 24-inch-wide
____	1	Scoop shovel
____	1	Pointed shovel
____	1	Spade
____	1	Trenching tool (folding)
____	3	ABC-type dry chemical fire extinguishers, 10-pound
____	3	Hand lights
____		Portable lights and generator for night extrications; match generator wattage to lights to be used.

Other Tools

Quantity and Item

____	2	Extension cords with ground, 12-gauge, 50-foot
____	1	Folding ladder, 12-gauge, 8-foot
____	1	Fiberglass pike pole, 8-foot
____	1	Pike pole, 36-inch closet hook
____	3	Mechanical jacks, handyman type, 7,000-pound-capacity
____	1	Hydraulic jack, 12-ton-capacity

Appendix B

Agricultural Operator Control Symbols

Newer farm tractors, combines, and other self-propelled units of agricultural equipment have universal operator control symbols so farmers can readily locate and operate the various controls (figures B.1, B.2, and B.3). These symbols can also assist rescue personnel in the event of an emergency. Being able to locate the fuel shutoff or operate a remote hydraulic system could contribute to a successful rescue.

Note: These symbols are generally not found on older agricultural equipment.

Signal Words

CAUTION is used for general reminders of good safety practices or to direct attention to unsafe practices. The color used for caution signals is yellow.

WARNING denotes a specific potential hazard. The color used for warning signals is orange.

DANGER denotes the most serious potential hazard. The color used for danger signals is red.

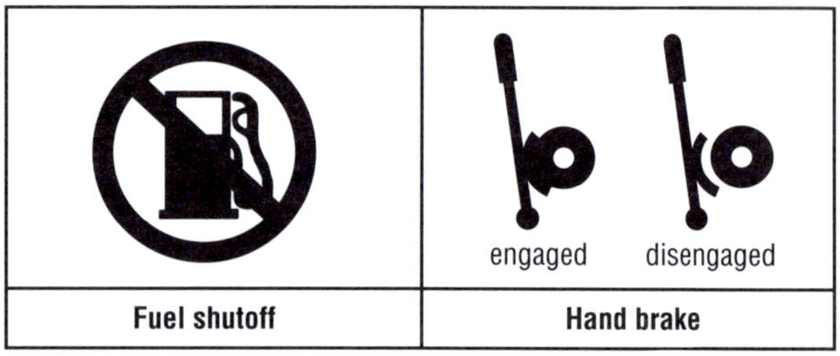

Figure B.1
Dashboard controls

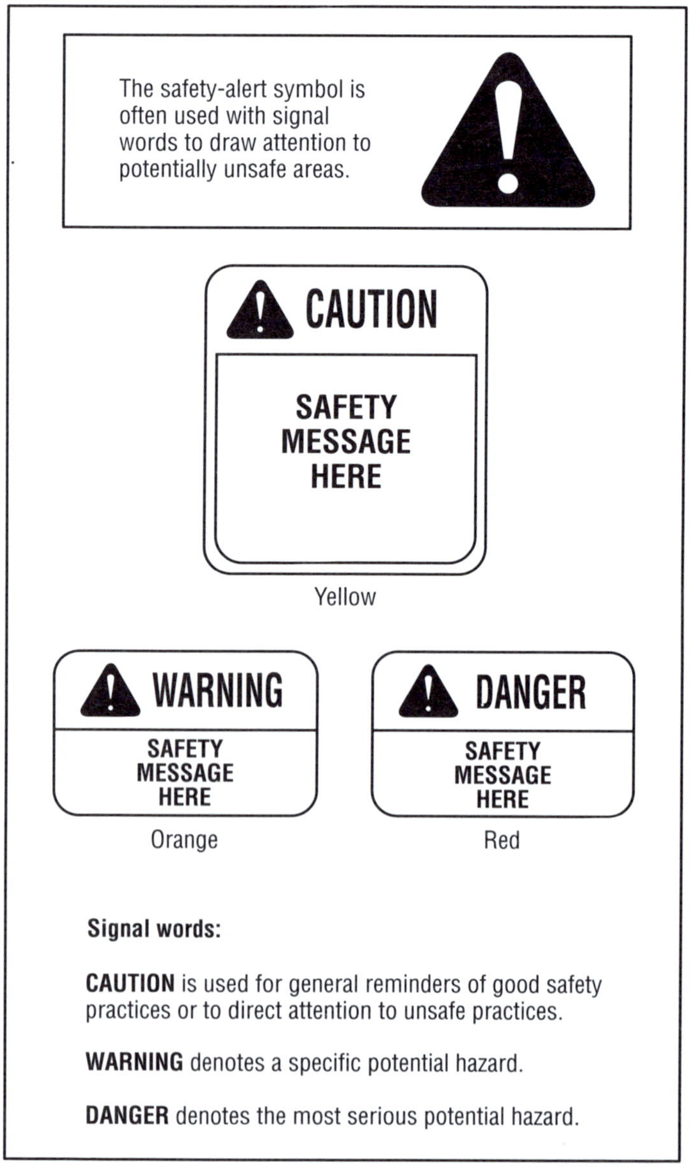

Figure B.2
Safety alert signals

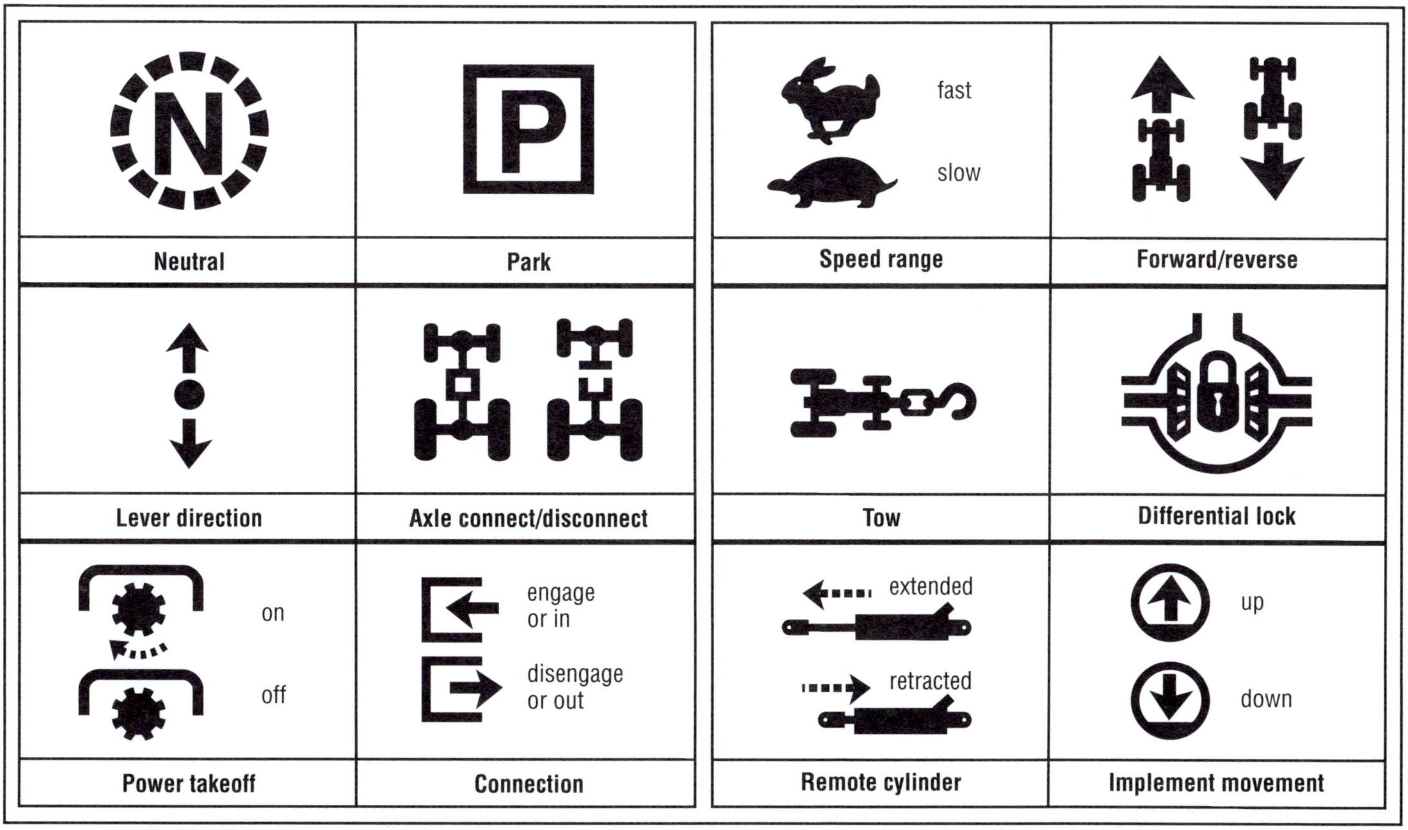

Figure B.3
Operational controls

Appendix C

Follow-up Training and Suggested Readings

Follow-up Training

Follow-up training is important. It should include practice runs to selected farm sites, tours of farm operations, and other opportunities to become better acquainted with various agricultural hazards.

Owners of agribusiness firms are usually very cooperative in providing emergency units with specialized instruction concerning their products or services. Most farm implement dealers and their service personnel can contribute effective and detailed instruction on the hazards associated with various pieces of farm equipment. Farm equipment mechanics have considerable insight into the best way to dismantle the more common machines.

Most agricultural chemical suppliers are also very willing to open their facilities for inspection and pre-emergency planning activities. Chemical suppliers are also a good source of information concerning safe handling of farm chemicals, as well as procedures to follow in the event of a spill or fire.

Another source of information on farm hazards and injuries is the state cooperative extension service or the extension safety specialist, who promotes farm safety and is knowledgeable of the circumstances surrounding most types of farm-related injuries and emergencies. The safety specialist may also have information on educational materials useful in training activities.

Cornell Farmedic Training Program

Agriculture is one of the most dangerous industries in America. Each year, needless injuries and deaths occur to farmers, family members, farm employees, and rescue personnel. Emergency responders often lack knowledge of the nature of farm machinery, chemicals, and farm structures; in addition, some rescue techniques may actually increase the risk to the victim and rescue personnel. Training and education in the methods of farm/rural rescue are essential for proper response and safety to the rescuer and patient. The main goal of the Cornell Farmedic Training Program is to provide rural fire/rescue responders with a systematic approach to farm rescue procedures that address the safety of both patients and responders.

The Cornell Farmedic Training Program includes an Instructor Training Program, "training the trainers" who will then train local responders through Provider Training Programs. Both trainings include classroom work, farm tours, demonstrations, and hands-on extrications using mannequins in real-life scenarios. An outreach program for the rural community at-large called "First on the Scene" is also available. The Instructor Course requires 18 hours of training over two days, and the Provider Course is 10 hours.

The Cornell Farmedic Training Program has been working in agricultural and rural incident response training for over two decades. Initial activities began in 1981 when the New York Farm Bureau, Empire Nine (a regional emergency services training program), Farm Family Insurance Company, and rural fire/rescue/EMS personnel got together to discuss the void in farm rescue education. Two troubling regional trends instigated this gathering of forces: (1) farm machinery extrications were often taking a long time—too long—to complete, and (2) rescuers were injured, some fatally, during rescues and fire-ground operations on farms. From the onset, a train-the-trainer approach was adopted as the strategy to educate rural rescuers to respond to fire and medical emergencies specific to farms. In the 1990s, the Cornell Farmedic Training Program partnered with Alfred State College and the New York State Department of Health. Since its inception, more than 28,000 rural fire/rescue/EMS, hospital personnel, and farm community members in 48 states and Canada have been trained how to respond to farm emergencies.

For further information about the Cornell Farmedic Training Program, contact:

Cornell Farmedic Training Program
777 Warren Road
Ithaca, New York 14850
Phone: (607) 257-9700
Phone, toll free: (800) 437-6010
Fax: (607) 257-5041
E-mail: farmedic@cornell.edu
Web site: www.farmedic.com

Sources of Information

Publications from NRAES

Before ordering, contact NRAES for current prices and shipping and handling charges. NRAES has over 130 publications; ask for a free catalog.

NRAES is a not-for-profit program dedicated to assisting land grant university faculty and others in increasing the public availability of research- and experienced-based knowledge. Read more about NRAES on page 78.

NRAES
Natural Resource, Agriculture, and Engineering Service
Cooperative Extension
PO Box 4557
Ithaca, New York 14852-4557
Phone: (607) 255-7654
Fax: (607) 254-8770
E-mail: nraes@cornell.edu
Web site: www.nraes.org

Extinguishing Fires in Silos and Hay Mows

34 pages • NRAES–18 • 2000 Revision • This book reviews the differences among conventional, oxygen-limiting, and modified oxygen-limiting silos, and the best methods to locate and extinguish fires in each. Hay mow fires are also discussed. Intended for firefighters, educators, silo dealers, and manufacturers.

Fire Control in Livestock Buildings

18 pages • NRAES–39 • 1989 • This publication helps farmers ensure that livestock buildings are constructed and equipped to minimize fire hazards. Topics discussed include the use of fire-retardant materials in construction, building management to limit the spread of fire, vent spacing, and early warning and automatic sprinkler systems.

First on the Scene

46 pages • NRAES–12 • 1989 • This publication provides detailed discussions and decision-and-action diagrams for specific farm accident scenarios involving machinery, storage facilities, chemicals, and electrocution. Those arriving first on the scene must understand the hazards and necessary reactions to enable a victim to survive.

On-Farm Agrichemical Handling Facilities

22 pages • NRAES–78 • 1995 • This publication discusses considerations a farmer should make regarding agrichemical storage, principal parts of the facility, storage environmental requirements, safety requirements, and storage alternatives.

Pesticides and Groundwater: A Guide for the Pesticide User

26 pages • NRAES–34 • 1995 • This publication is written for pesticide users and rural residents concerned about protecting groundwater. Included: a 10-page table of U.S. EPA drinking water contaminants found in pesticide products.

Used Farm Equipment: Assessing Quality, Safety, and Economics

34 pages • NRAES–25 • 1987 • This handbook shows the buyer how to inspect machinery for the reliability of its components and the quality of its safety features. The economics of owning and operating used machinery are covered, and methods of acquiring equipment are discussed.

Resources from Purdue University

Farm Family Safety and Health Workshop, Leader's Guide. 1996. This 95-page publication is designed to provide local leaders the planning tools needed to organize a farm safety and health workshop, including topics on emergency response. Available from the:

Agricultural Safety and Health Program
Department of Agricultural and Biological Engineering
Purdue University
1146 ABE Building
West Lafayette, Indiana 47907-1146
Phone: (765) 494-1162
Fax: (765) 496-1115
Web site: http://abe.www.ecn.purdue.edu/abe

A number of publications are available on-line from Purdue University Cooperative Extension at <www.agcom.purdue.edu/AgCom/Pubs/ageng.htm>. Topics include agricultural air quality, agricultural chemicals, and farm safety and health.

Publications from ASAE

ASAE—the Society for Engineering in Agricultural, Food, and Biological Systems—is a not-for-profit professional and technical organization of members worldwide interested in engineering knowledge and technology for food and agriculture, associated industries, and related resources.

ASAE
2950 Niles Road
St. Joseph, Michigan 49085-9659
Phone: (616) 429-0300
Fax: (616) 429-3852
E-mail: HQ@ASAE.ORG
Web site: WWW.ASAE.ORG

Journal of Agricultural Safety and Health

An interdisciplinary, refereed journal designed to foster identification and discussion of the issues of agricultural safety and health. Published quarterly. Contact ASAE for more information and for subscription rates.

Safety and Health for Production Agriculture

Topics in this book include production agriculture, safety and health hazards, injury statistics, fundamentals of occupational safety and health, hazard and injury prevention and control, voluntary safety behavior change, regulations to require behavior change, and engineering for hazard and injury prevention and control. 1992.

Publications from AAOS

Rural Rescue and Emergency Care and *Basic Rescue and Emergency Care* were published by the American Academy of Orthopaedic Surgeons (AAOS). Check with bookstores or libraries for these two publications. They are no longer available from AAOS and may be out of print.

Rural Rescue and Emergency Care

Rural Rescue and Emergency Care is designed for rescuers responding to agricultural/rural incidents. The publication discusses pre-incident planning and rescuer preparation; emergencies involving farm machinery, farm structures, and agricultural chemicals; and some of the hazards rescuers may be exposed to. 1993.

Basic Rescue and Emergency Care

This publication discusses the five distinct elements of the rescue operation: pre-planning, plus the four phases of the rescue itself: *locating* the subject of the rescue operation, *accessing* the patient safely, *stabilizing* patients during an extended incident, and *transporting* the patient safely. 1990.

Resources from the Emergency Medical Services for Children (EMSC)

The organization Emergency Medical Services for Children (EMSC) has a variety of resources on enhancing emergency care for children. It includes information on upgrading emergency vehicles to deal with pediatric trauma and proposed guidelines for pediatric emergency care. Contact:

EMSC Clearinghouse
2070 Chain Bridge Road, Suite 450
Vienna, Virginia 22182-2536
Phone: (703) 902-1203
Fax: (703) 821-2098
E-mail: EMSC@CIRCSOL.COM
Web site: WWW.EMS-C.ORG

Resources from NIOSH

The following publications are available from the National Institute for Occupational Safety and Health (NIOSH): *Preventing Entrapment and Suffocations Caused by the Unstable Surfaces of Stored Grain* (1987), *Preventing Grain Auger Electrocutions* (1986), and *Safe Grain and Silage Handling* (1995).

National Institute for Occupational Safety and Health (NIOSH)
NIOSH Publications
4676 Columbia Parkway, Mail Stop C-13
Cincinnati, Ohio 45226-1998
Phone, toll fee: (800) 356-4674 [(800) 35-NIOSH]
Phone, outside the U.S.: (513) 533-8471
Fax: (513) 533-8573
E-mail: PUBSTAFT@CDC.GOV
Web site, NIOSH home page: WWW.CDC.GOV/NIOSH
Web site, publications: WWW.CDC.GOV/NIOSH/PUBS.HTML

Additional Resources

On-Line Resources

The following World Wide Web sites furnish a variety of information that is relevant to providing emergency services in rural communities and agricultural settings.

URLs (or uniform resource locators) are given for each organization's web site. URLs change frequently; if a web site listed in this section is out-of-date, search the Internet for the name of the organization and the most current URL.

American Lung Association

Web site: <www.lungusa.org>. Information about lung diseases. Some information is agriculture-related.

Center for Rural Emergency Medicine (CREM)

Web site: <www.hsc.wvu.edu/crem>. Education, research, service, gopher archive, and data on the delivery of emergency and trauma care in rural areas.

Cooperative Extension Service, Federal

Web site: <www.ace.orst.edu/info/nptn/exten.htm>. Access to resources at land grant universities across the country.

Children's Safety Network (CSN), Education Development Center

Web site: <www.edc.org/HHD/csn>. (Note: This URL is case-sensitive. Use capital letters for "HHD.") Project highlights, CSN site descriptions, and links to other injury-prevention sites.

Emergency Medical Services for Children (EMSC)

Web site: <www.ems-c.org>. Publications, data, and resources on enhancing emergency medical services for children.

Farm Safety 4 Just Kids

Web site: <www.fs4jk.org>. Information for parents, kids, and chapters interested in childhood agricultural safety.

Indiana Rural Safety and Health Council (IRSCH)

Web site: <www.farmsafety.org>. Available resources on health and safety for rural and farm families. Additional links to other injury-prevention sites. Useful educational pages can be downloaded.

Minnesota Farm Safety and Health

Web site: <www.bae.umn.edu/~fs>. Activities in Minnesota and links to national and international occupational safety and health sites.

National Ag Safety Database

Web site: <www.cdc.gov/niosh/nasd/nasdhome.html>. Extensive compendium of safety and health resources, including videos, directories, publications, and abstracts.

National Center for Farmworker Health (NCFH)

Web site: <www.ncfh.org>. Information resources and special projects of NCFH with migrant and seasonal farm workers.

National Children's Center for Rural and Agricultural Health and Safety (NCCRAHS)

Web site: <research.marshfieldclinic.org/children/>. (Note: No "www" in URL.) Provides resources, research, and education to enhance the health and safety of all children exposed to agricultural hazards in the U.S., with a focus on injury prevention.

National Farm Medicine Center

Web site: <www.marshmed.org/nfmc>. Resources dedicated to improving the quality of life on farms and ranches. Links to National Children's Center for Rural and Agricultural Health and Safety.

National Highway and Traffic Safety Administration (NHTSA)

Web site: <www.nhtsa.dot.gov>. Events, "What's New?" feature, and links to other transportation sites.

National Institute for Occupational Safety and Health (NIOSH)

Web site: <www.cdc.gov/niosh/homepage.html>. Information, services, and links to other occupational web sites.

National SAFE KIDS Campaign

Web site: <www.safekids.org>. Fact sheets available on a variety of child-injury topics.

Ohio Agricultural Safety Page

Web site: <www2.ag.ohio-state.edu/~agsafety>. Listing of publications, services, and products available from the Ohio State University Extension.

Purdue Agricultural Safety and Health Program

Web site: <pasture.ecn.purdue.edu/~agsafety>. (Note: No "www" in URL.) Extensive listing of agricultural safety and health resources, including links to other helpful sites.

Rural Information Center Health Services (RICHS)

Web site: <www.nal.usda.gov/ric/richs>. Upcoming conferences, funding sources, publications, and policy and research issues.

University of California (UC) Agricultural Health and Safety Center

Web site: <agcenter.ucdavis.edu/agcenter>. (Note: No "www" in URL.) NIOSH center abstracts, projects, newspaper clippings, information, and links to other sites.

Don't Go with the Flow! Commercial Grain Storage Resource Instructional Package

This package—*Don't Go with the Flow! Commercial Grain Storage Resource Instructional Package*—includes a 28-minute video, instructor's guide, overhead masters, and employee poster designed to train local and plant first-response personnel on appropriate rescue procedures for large commercial grain storage facilities. 1998. Available from:

National Grain and Feed Association (NGFA)
1201 New York Avenue NW, Suite 830
Washington, DC 20005-3917
Phone: (202) 289-0873
Fax: (202) 289-5388
E-mail: NGFA@NGFA.ORG
Web site: WWW.NGFA.ORG

Farm and Ranch Safety Management

Farm and Ranch Safety Management is a vital management tool for creating a safe working environment within agricultural enterprises. This publication (and an instructor guide, student guide, and slide set) are available from John Deere Publishing. Call for current pricing. Ask for a free catalog.

John Deere Publishing
Department 373
5440 Corporate Park Drive
Davenport, Iowa 52807
Phone, toll free: (800) 522-7448
Fax: (319) 355-3690
E-mail: LL40294@DEERE.COM
Web site: WWW.DEERE.COM

Fire Protection Guide to Hazardous Materials

The publication *Fire Protection Guide to Hazardous Materials* includes four documents from the NFPA (National Fire Protection Agency) that classify materials so that response personnel can safely handle emergencies such as fires, accidental releases, and highway accidents. This and many other resources are available from NFPA:
Phone, customer sales: (800) 344-3555
Phone, outside the U.S.: (508) 895-8300
E-mail: CUSTSERV@NFPA.ORG
Web site: WWW.NFPA.ORG

1996 North American Emergency Response Guidebook

The publication *1996 North American Emergency Response Guidebook* was developed by the U.S. Department of Transportation (DOT) to aid first responders, when there are transportation incidents involving hazardous materials. Catalog number 050-000-00561-5. Available from:

Superintendent of Documents
Government Printing Office
Phone: (202) 512-1800
Fax: (202) 512-2250
Web site: WWW.ACCESS.GPO.GOV/SU_DOCS/SALE.HTML

DOT has announced plans to produce an updated version: *2000 North American Emergency Response Guidebook.*

Out-of-Print References

The following two references are out-of-print but may still be available from libraries or used booksellers.

Recognition and Management of Pesticide Poisonings. A 120-page book designed to help health professionals deal with persons exposed to pesticides. Originally published by the U.S. Government Printing Office, Washington, DC.

Preplanning and Guidelines for Handling Agricultural Chemical Fires. General guidelines to reduce hazards when dealing with agricultural chemical fires. Originally published by the National Agricultural Chemicals Association, Washington, DC.

Emergency Telephone Numbers

For Rescue Personnel's Use.
Fill in This Page. Keep It Handy in Case of an Emergency.

Police, Fire, Hospital, Other Emergency Contacts

Air ambulance ____________________

Ambulance ____________________

Fire department ____________________

Hazardous response team ____________________

Hospital ____________________

Paramedic unit ____________________

Police (local) ____________________

Police (state) or sheriff ____________________

Silo or high-rise rescue team ____________________

Towing service ____________________

Other emergency contacts ____________________

____

____

____

____

____

Other Contacts

Local Farm Machinery Dealers and Service Managers

1. Business ____________________

   Home ____________________

2. Business ____________________

   Home ____________________

Hydraulic Injuries (see pages 15–18)

John Deere and Company 1-800-822-8262

____

Extrications

Mechanics with extrication experience ____________________

____

____

Chemical and Pesticide Emergencies (see chapter 4)

CHEMTREC 1-800-424-9300

____

Poison control center ____________________

About NRAES

NRAES, the Natural Resource, Agriculture, and Engineering Service, is a not-for-profit program dedicated to assisting land grant university faculty and others in increasing the public availability of research- and experienced-based knowledge. NRAES is sponsored by fourteen land grant universities in the eastern United States (see map below). We receive administrative support from Cornell University, the host university.

When you buy books from NRAES, you are helping to improve the accessibility of land grant university knowledge. While 15% of NRAES' annual income is provided by member universities, the funds to publish new books and coordinate new conferences come from our customers through book sales, conference registrations, and occasional project-specific grants.

NRAES publishes practical books of interest to fruit and vegetable growers, landscapers, dairy and livestock producers, natural resource managers, SWCD (soil and water conservation district) staff, consumers, landowners, and professionals interested in agricultural waste management and composting. NRAES books are used in cooperative extension programs, in college courses, as management guides, and for self-directed learning.

NRAES publishes two types of books: peer-reviewed books and conference proceedings. Our peer-reviewed books are evaluated prior to publication for technical accuracy and usefulness to the intended audience. The reviewers may include university faculty, extension educators, potential users, and interested persons from government and agribusiness. Conference proceedings are not peer-reviewed. However, the authors of papers presented at NRAES-sponsored conferences are chosen for their recognized expertise. NRAES also distributes some videos related to waste management and fruit and vegetable production.

NRAES was started in 1974 and was originally known as the Northeast Regional Agricultural Engineering Service. In 1987, with encouragement from member university extension directors, NRAES began offering its services to faculty from all disciplines at member universities. In 1998, Virginia Polytechnic Institute and State University joined the original thirteen member universities. Our name was changed in 1998 to reflect the expansion of NRAES beyond the Northeast and the broadening scope of our books and conferences.

Contact us for more information or a free book catalog:

NRAES
Natural Resource, Agriculture, and Engineering Service
Cooperative Extension
PO Box 4557
Ithaca, New York 14852-4557
Phone: (607) 255-7654 E-mail: NRAES@CORNELL.EDU
Fax: (607) 254-8770 Web site: WWW.NRAES.ORG

Marty Sailus, NRAES Director
Jeffrey S. Popow, NRAES Managing Editor

NRAES Member Universities

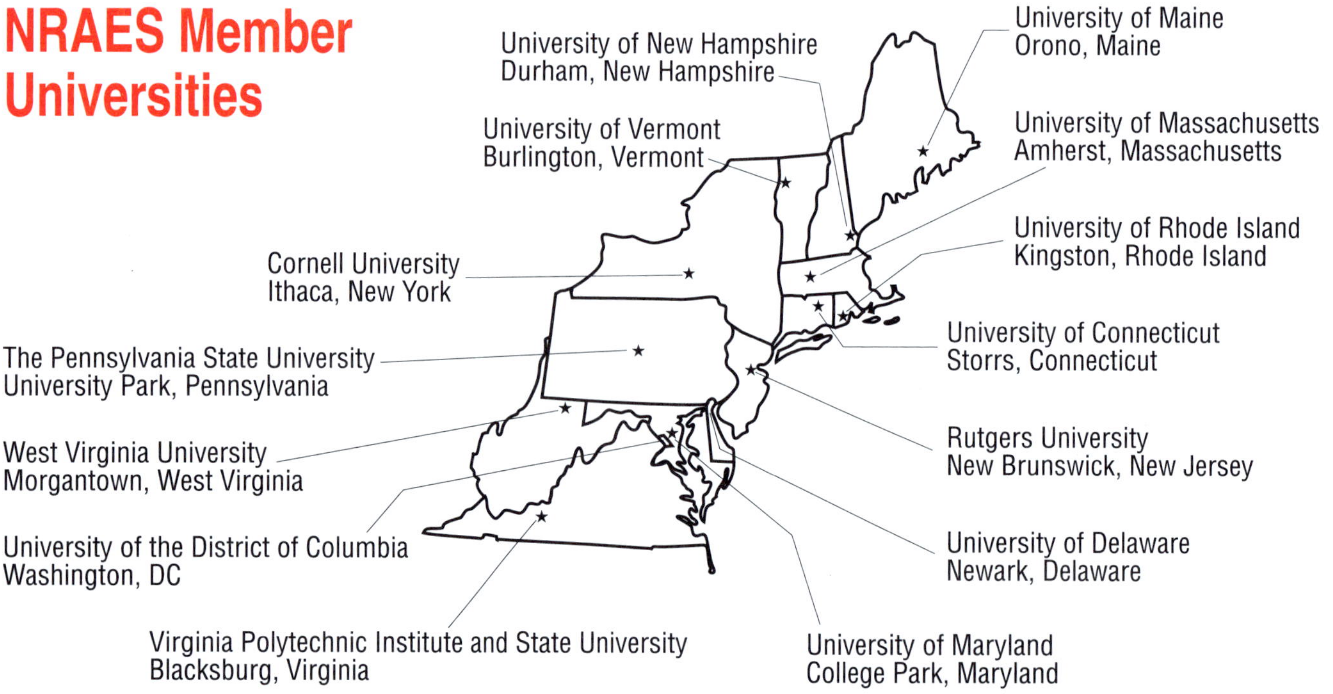